ENCYCLOPAEDIA OF PHOTOSYNTHESIS

ENCYCLOPAEDIA OF PHOTOSYNTHESIS

Vol. 1

Udai Veer

ANMOL PUBLICATIONS PVT. LTD.
NEW DELHI - 110 002 (INDIA)

ANMOL PUBLICATIONS PVT. LTD.
H.O.: 4374/4B, Ansari Road, Darya Ganj,
New Delhi-110 002 (India)
Ph.: 23278000, 23261597
B.O.: No. 1015, Ist Main Road, BSK IIIrd Stage
IIIrd Phase, IIIrd Block,
Bangalore - 560 085 (India)
Visit us at: www.anmolpublications.com

Encyclopaedia of Photosynthesis

First Adition, 2008

ISBN 978-81-261-3474-8 (Set)

PRINTED IN INDIA

Printed at Mehra Offset Press, Delhi.

Contents

Preface

Phtosynthesis is the process, by which plants make their food in their green leaves with the help of carbon dioxide, minerals, and water; in the presence of light. In this process oxygen is given out, which is very useful in the respiration. This is the process, in which light energy is captured in green leaves for the chemical process, which makes it most important and fascinating of biological processes. By the absorption of blue light, chlorophyll emits only the useful red fluorescence. The molecule goes from its second to first excited singlet state by a radiationless process of high probability. The carbohydrate made in the process of photosynthesis is synthesised by animals and plants both. They synthesise fats and proteins from the carbohydrate; thus glucose is a basic energy source for all living organisms.

Photosynthesis fills all of the food requirements directly or indirectly. The energy stored in petroleum, natural gas and coal, has all come from the sun via photosynthesis, as does the energy in firewood, which is a major fuel in many parts of the world. All of the biological energy needs are met by the plant kingdom, either directly or through herbivorous animals. One of the carbohydrates resulting from photosynthesis is cellulose, which makes up the bulk of dry wood and other plant material. When the wood is burnt the cellulose is converted back to carbon dioxide and releases the stored energy as heat. Burning fuel is basically the same oxidation process that occurs in living bodies. It liberates the energy of stored sunlight in a useful form, and returns carbon dioxide to the atmosphere.

Energy from burning biomass is important in many parts of the world. In developing countries, firewood continues to be critical for survival. Ethanol (grain alcohol) produced from sugar and starch by fermentation, is a major automobile fuel in Brazil and is added to gasoline in some parts of the United States in order to help reduce emission of harmful pollutants.

Photosynthesis by plants removes carbon dioxide from the atmosphere and replaces it with oxygen. Thus, it tends to ameliorate the effects of carbon dioxide released by the burning of the fossil fuels. Photosynthetic pigments come in a huge variety. There are different types of chlorophyll, carotenoids and phycobilins, differing from each other, in their precise chemical structure.

Generally pigments are bound by proteins, which provide the pigment molecules with appropriate positioning and orientation, with respect to each other.

This eminent work: *Encyclopaedia of Photosynthesis* has a very realistic and systematic feature, which would be able to make a distinct impression on the study of photosynthesis. Expectedly, each and every reader will get full satisfaction by studying this book.

Editor

Introduction

Photosynthesis, process by which green plants and certain other organisms use the energy of light to convert carbon dioxide and water into the simple sugar glucose. In so doing, photosynthesis provides the basic energy source for virtually all organisms. An extremely important by-product of photosynthesis is oxygen, on which most organisms depend.

Photosynthesis occurs in green plants, seaweeds, algae, and certain bacteria. These organisms are veritable sugar factories, producing millions of new glucose molecules per second. Plants use much of this glucose, a carbohydrate, as an energy source to build leaves, flowers, fruits, and seeds. They also convert glucose to cellulose, the structural material used in their cell walls. Most plants produce more glucose than they use, however, and they store it in the form of starch and other carbohydrates in roots, stems, and leaves. The plants can then draw on these reserves for extra energy or building materials. Each year, photosynthesising organisms produce about 170 billion metric tons of extra carbohydrates, about 30 metric tons for every person on earth.

Photosynthesis has far-reaching implications. Like plants,

humans and other animals depend on glucose as an energy source, but they are unable to produce it on their own and must rely ultimately on the glucose produced by plants. Moreover, the oxygen humans and other animals breathe is the oxygen released during photosynthesis.

Humans are also dependent on ancient products of photosynthesis, known as fossil fuels, for supplying most of our modern industrial energy. These fossil fuels, including natural gas, coal, and petroleum, are composed of a complex mix of hydrocarbons, the remains of organisms that relied on photosynthesis millions of years ago. Thus, virtually all life on earth, directly or indirectly, depends on photosynthesis as a source of food, energy, and oxygen, making it one of the most important bio-chemical processes known.

Where Photosynthesis Occurs - ?

Plant photosynthesis occurs in leaves and green stems within specialised cell structures called chloroplasts. One plant leaf is composed of tens of thousands of cells, and each cell contains 40 to 50 chloroplasts. The chloroplast, an oval-shaped structure, is divided by membranes into numerous disk-shaped compartments. These disklike compartments, called thylakoids, are arranged vertically in the chloroplast like a stack of plates or pancakes. A stack of thylakoids is called a granum (plural, *grana*); the grana lie suspended in a fluid known as stroma.

Embedded in the membranes of the thylakoids are hundreds of molecules of chlorophyll, a light-trapping pigment required for photosynthesis. Additional light-trapping pigments, enzymes (organic substances that speed up chemical reactions), and other molecules needed for photosynthesis are also located within the thylakoid membranes. The pigments and enzymes are arranged in two types of units, Photosystem I and Photosystem II. Because a chloroplast may have dozens of thylakoids, and each thylakoid may contain thousands of photosystems, each chloroplast will contain millions of pigment molecules.

How Photosynthesis Works - ?

Photosynthesis is a very complex process, and for the sake of convenience and ease of understanding, plant biologists divide it into two stages. In the first stage, the light-dependent reaction, the chloroplast traps light energy and converts it into chemical energy contained in nicotinamide adenine dinucleotide phosphate (NADPH) and adenosine triphosphate (ATP), two molecules used in the second stage of photosynthesis. In the second stage, called the light-independent reaction (formerly called the dark reaction), NADPH provides the hydrogen atoms that help form glucose, and ATP provides the energy for this and other reactions used to synthesise glucose. These two stages reflect the literal meaning of the term *photosynthesis,* to build with light.

Light-dependent Reaction

Photosynthesis relies on flows of energy and electrons initiated by light energy. Electrons are minute particles that travel in a specific orbit around the nuclei of atoms and carry a small electrical charge. Light energy causes the electrons in chlorophyll and other light-trapping pigments to boost up and out of their orbit; the electrons instantly fall back into place, releasing resonance energy, or vibrating energy, as they go, all in millionths of a second. Chlorophyll and the other pigments are clustered next to one another in the photosystems, and the vibrating energy passes rapidly from one chlorophyll or pigment molecule to the next, like the transfer of energy in billiard balls.

Light contains many colours, each with a defined range of wavelengths measured in nanometres, or billionths of a metre. Certain red and blue wavelengths of light are the most effective in photosynthesis because they have exactly the right amount of energy to energise, or excite, chlorophyll electrons and boost them out of their orbits to a higher energy level. Other pigments, called accessory pigments, enhance the light-absorption capacity of the leaf by capturing a broader spectrum of blue and red wavelengths, along with yellow and orange wavelengths. None of the photosynthetic pigments absorb green light; as a result, green wavelengths are reflected, which is why plants appear green.

Photosynthesis begins when light strikes Photosystem I pigments and excites their electrons. The energy passes rapidly from molecule to molecule until it reaches a special chlorophyll molecule called P700, so named because it absorbs light in the red region of the spectrum at wavelengths of 700 nanometres.

Until this point, only energy has moved from molecule to molecule; now electrons themselves transfer between molecules. P700 uses the energy of the excited electrons to boost its own electrons to an energy level that enables an adjoining electron acceptor molecule to capture them. The electrons are then passed down a chain of carrier molecules, called an electron transport chain.

The electrons are passed from one carrier molecule to another in a downhill direction, like individuals in a bucket brigade passing water from the top of a hill to the bottom. Each electron carrier is at a lower energy level than the one before it, and the result is that electrons release energy as they move down the chain. At the end of the electron transport chain lies the molecule nicotine adenine dinucleotide ($NADP^+$). Using the energy released by the flow of electrons, two electrons from the electron transport chain combine with a hydrogen ion and $NADP^+$ to form NADPH.

When P700 transfers its electrons to the electron acceptor, it becomes deficient in electrons. Before it can function again, it must be replenished with new electrons. Photosystem II accomplishes this task. As in Photosystem I, light energy activates electrons of the Photosystem II pigments. These pigments transfer the energy of their excited electrons to a special Photosystem II chlorophyll molecule, P680, that absorbs light best in the red region at 680 nanometres. Just as in Photosystem I, energy is transferred among pigment molecules and is then directed to the P680 chlorophyll, where the energy is used to transfer electrons from P680 to its adjoining electron acceptor molecule.

From the Photosystem II electron acceptor, the electrons are passed through a different electron transport chain. As they pass along the cascade of electron carrier molecules, the electrons give up some of their energy to fuel the production of ATP, formed by the addition of one phosphorus atom to adenosine diphosphate (ADP). Eventually, the electron transport carrier molecules deliver the Photosystem II electrons to Photosystem I, which uses them to maintain the flow of electrons to P700, thus restoring its function.

P680 in Photosystem II is now electron deficient because it has donated electrons to P700 in Photosystem I. P680 electrons are replenished by the water that has been absorbed by the plant roots and transported to the chloroplasts in the leaves. The movement of electrons in Photosystems I and II and the action of an enzyme split the water into oxygen, hydrogen ions, and electrons. The electrons from water flow to Photosystem II, replacing the electrons lost by P680. Some of the hydrogen ions may be used to produce NADPH at the end of the electron transport chain, and the oxygen from the water diffuses out of the chloroplast and is released into the atmosphere through pores in the leaf.

The transfer of electrons in a step-by-step fashion in Photosystems I and II releases energy and heat slowly, thus protecting the chloroplast and cell from a harmful temperature increase. It also provides time for the plant to form NADPH and ATP. In the words of American biochemist and Nobel laureate Albert Szent-Gyorgyi, "What drives life is thus a little electric current, set up by the sunshine."

Light-independent Reaction

The chemical energy required for the light-independent reaction is supplied by the ATP and NADPH molecules produced in the light-dependent reaction. The light-independent reaction is cyclic, that is, it begins with a molecule that must be regenerated at the end of the reaction in order for the process to continue. Termed the Calvin cycle after the American chemist Melvin Calvin who discovered it, the light-independent reactions use the electrons and hydrogen ions associated with NADPH and the phosphorus associated with ATP to produce glucose. These reactions occur in the stroma, the fluid in the chloroplast surrounding the thylakoids, and each step is controlled by a different enzyme.

The light-independent reaction requires the presence of carbon dioxide molecules, which enter the plant through pores in the leaf, diffuse through the cell to the chloroplast, and disperse in the stroma. The light-independent reaction begins in the stroma when these carbon dioxide molecules link to sugar molecules called ribulose bisphosphate (RuBP) in a process known as carbon fixation.

With the help of an enzyme, six molecules of carbon dioxide bond to six molecules of RuBP to create six new molecules. Several intermediate steps, which require ATP, NADPH, and additional enzymes, rearrange the position of the carbon, hydrogen, and oxygen atoms in these six molecules, and when the reactions are complete, one new molecule of glucose has been constructed and five molecules of RuBP have been reconstructed. This process occurs repeatedly in each chloroplast as long as carbon dioxide, ATP, and NADPH are available. The thousands of glucose molecules produced in this reaction are processed by the plant to produce energy in the process known as aerobic respiration, used as structural materials, or stored. The regenerated RuBP is used to start the Calvin cycle all over again.

Photosynthesis Variations

A majority of plants use these steps in photosynthesis. Plants such as corn and crabgrass that have evolved in hot, dry environments, however, must overcome certain obstacles to photosynthesis. On hot days, they partially close the pores in their leaves to prevent the escape of water. With the pores only slightly open, adequate amounts of carbon dioxide cannot enter the leaf, and the Calvin cycle comes to a halt. To get around this problem, certain hot-weather plants have developed a way to keep carbon dioxide flowing to the stroma without capturing it directly from the air.

They open their pores slightly, take in carbon dioxide, and transport it deep within the leaves. Here they stockpile it in a chemical form that releases the carbon dioxide slowly and steadily into the Calvin cycle. With this system, these plants can continue photosynthesis on hot days, even with their pores almost completely closed. A field of corn thus remains green on blistering days when neighbouring plants wither, and crabgrass thrives in lawns browned by the summer sun.

Bacteria lack chloroplasts, and instead use structures called chromatophores—membranes formed by numerous foldings of the plasma membrane, the membrane surrounding the fluid, or cytoplasm, that fills the cell. The chromatophores house thylakoids similar to plant thylakoids, which in some bacteria contain chlorophyll. For these bacteria, the process of photosynthesis is similar to that of plants, algae, and seaweed. Many of these chlorophyll-containing bacteria are abundant in oceans, lakes, and rivers, and the oxygen they release dissolves in the water and enables fish and other aquatic organisms to survive.

Certain archaebacteria, members of a group of primitive bacteria-like organisms, carry out photosynthesis in a different manner. The mud-dwelling green sulphur and purple sulphur archaebacteria use hydrogen sulphide instead of water in photosynthesis. These archaebacteria release rather than oxygen, which, along with hydrogen sulphide, imparts the rotten egg smell to mudflats. Halobacteria, archaebacteria found in the salt flats of deserts, rely on the pigment bacteriorhodopsin instead of chlorophyll for photosynthesis. These archaebacteria do not carry out the complete process of photosynthesis; although they produce ATP in a process similar to the light-dependent reaction and use it for energy, they do not produce glucose. Halobacteria are among the most ancient organisms, and may have been the starting point for the evolution of photosynthesis.

While it may seem that we understand photosynthesis in detail, decades of experiments have given us only a partial understanding of this important process. A more thorough understanding of the details of photosynthesis may pave the way for development of crops that are more efficient at using the sun's energy, producing food for increasingly bountiful harvests.

Factors that Influence Photosynthesis

Environmental Factors Affecting Photosynthesis: In simple terms photosynthesis is a plants ability to derive energy from light. In a more technical sense photosynthesis can be described as a process by which autotrophic organisms utilise chloroplast pigments to synthesise complex carbohydrates. This process is only active in most photosynthetic organisms in light wavelengths between 400-700 angstroms (visible light spectrum). There are many factors that influence the effectiveness and rate by which plants and some bacterial organisms achieve this process.

Role of Light

Light is why leafs exist: "to capture a ray of light in essence is to touch the energy that drives all life", a wise man once said. The beginning process of photosynthesis is a plants access to light. Once light hits a plants leaves the photosynthetic process is started, as chlorophyll in the leaves begins to transform the energy received from the light. With out light, plants can not survive, so light availability is arguably one of the most, if not the most, important factor for photosynthesis. (Farabee: 2001).

Despite this obvious need for light however, light intensity can in fact be damaging to the photosynthetic process. Varying wavelengths of light can cause this adverse affect including ultraviolet wavelengths, light in the visible spectrum, and an interaction of these two wavelengths. If a leaf or other photosynthetic organism is exposed to high light intensity for a relatively long amount of time, it would result in a destruction of the photosynthetic cells within an organism. If this exposure continues for a long enough amount of time it can lead to the death of the organism.

Different plant types exhibit different tolerance levels to light intensity. For example a shade plant is quite a bit less tolerant to light intensity then a sun plant would be in that damaging effects would be seen sooner in the shade plants (Powles:1984). The term for light intensity damage to a plant is photo-inhibition. While this toxic effect is potentially deadly, if a plant is only exposed for a short duration of time, it can recover and repair the damages done to it once it is removed from the high light intensity. So while light is a very important factor in photosynthesis, too much of it, or rather too high of an intensity level, can become detrimental to the plant.

Role of Temperature

Temperature plays one of the most important roles in the rate and ability of a plant to photosynthesise effectively. There are several components in the process that are affected through a given temperature range: enzymes and water availability are two such components. In a general sense there is a positive correlation between change in temperature and photosynthesis. This means that as temperatures increase photosynthetic rates increase and as temperature decreases the photosynthetic rate also decreases. These increases and decreases however do occur in a certain temperature range and if a temperature is either too high or too low photosynthesis will cease to occur (plants survival at stake due to extreme temperatures). The range of temperature change that is acceptable for a plant varies from species to species.

A study conducted in New York which related temperature

differences to photosynthetic rates of red spruce, discovered that soil temperature actually affects a plant more than air temperature does.

This study concluded that not only was net photosynthesis more sensitive to changes in soil temperature, but that temperature differences in the fall were almost twice as sensitive as temperature changes in the spring (Schwarz: 1997).

Temperature Affecting Enzymes

Temperature directly affects the ability of enzymes and thus enzyme processes that take place in the leaf of a plant. In recent research, the effects of both high temperature and low temperature have similar outcomes, but these results occur via different path ways.

High temperature affects the structure of an enzyme. Although high temperature related to activity can be beneficial, there is a point on every enzymes activity curve where activity rapidly decreases. This decrease in activity is due to the heat energy associated with high temperatures.

Denaturation is the unravelling of a protein's (all enzymes are proteins, but not all proteins are enzymes) tertiary structure, or its double helical- three dimensional conformation. This is in its most basic sense, changing the shape of an enzyme. More over, the shape of an enzyme is its primary tool for catalysing reactions that would otherwise be too energy consuming to process (enzymes lower the energy of activation of a particular reaction, i.e. they lower the amount of energy necessary to activate a reaction).

When this shape is distorted by damaging heat, the enzyme can no longer function as an effective unit in the process of photosynthesis.

There are many enzymes that play key roles in the synthesis of carbohydrates by photosynthesis. Eliminating just one will likely halt production until conditions are once again favourable.

Temperature of Water and Water Availability

In the case of waters affect on photosynthesis, water availability has a greater impact on photosynthetic rates then the temperature of water does.

The only great impact felt from water temperature is when extreme hot and cold temperatures are reached (freezing and boiling). At either of these points plants either can't take up the solid water or are killed by the boiling water, in which cases photosynthesis does not apply. Otherwise, a slight variation in overall temperature of the water seems to have little to no affect on the photosynthetic rate.

Water availabilities affect on photosynthesis has a lot to do with the plants transpiration rates as well. In a period of water deficiency plants will force a closure of their stomata so as to decrease loss of water through transpiration, thus slowing down the photosynthetic rate as well. A decrease in water availability can actually increase water use efficiency of the plant since more of the water is being retained with in the plant systems (Martin, 1992). With a high water availability a plant is actively transpiring for longer time durations and therefore will have high photosynthetic rates as well. In a general sense however photosynthesis will increase with increasing water availability (to a certain point) and decrease with decreasing water availability.

Altitude and Carbon Dioxide

Carbon dioxide (CO_2) uptake by a plant adds to the photosynthetic process by providing a source of energy to the plant. Through carbon dioxide sequestration a plant is able to convert some of the CO_2 into sugars which can then be used as energy. Altitude is an environmental factor that affects the process by which some plants chemically respire and photosynthesise and can also affect a plants CO_2 intake. One instance is demonstrated in the occurrence of the way two plants at different altitudes fix CO_2. Some research suggests that plants at higher altitudes fix CO_2 directly from the atmosphere and from their own respiration. Plants at lower altitudes differed in that they tend not to fix CO_2 from their own respiration.

Consequently, plants at higher altitudes may have access to higher levels of CO_2 and higher levels of photons. This is because there are less canopy producing plants at higher altitudes due to the lower amounts of soil and water. This would lead one to believe that there would be higher levels of photosynthesis at higher altitudes due to possibly higher CO_2 fixation, but this does not seem to be the case. As altitude increases it appears as though effectiveness of plant photosynthesis is not affected, but this still has not been thoroughly researched. There is some research that relates altitude to enzyme activity and the final products of photosynthesis, but the rate at which these products are produced is not effected (Kumar, 2006).

Significance of Photosynthesis

Animals and plants both synthesise fats and proteins from carbohydrates; thus glucose is a basic energy source for all living organisms. The oxygen released (with water vapour, in transpiration) as a photosynthetic by-product, principally of phytoplankton, provides most of the atmospheric oxygen vital to respiration in plants and animals, and animals in turn produce carbon dioxide necessary to plants. Photosynthesis can therefore be considered the ultimate source of life for nearly all plants and animals by providing the source of energy that drives all their metabolic processes.

Photosynthesis is arguably the most important biological process on earth. By liberating oxygen and consuming carbon dioxide, it has transformed the world into the hospitable environment we know today. Directly or indirectly, photosynthesis fills all of our food requirements and many of our needs for fibre and building materials. The energy stored in petroleum, natural gas and coal all came from the sun via photosynthesis, as does the energy in firewood, which is a major fuel in many parts of the world. This being the case, scientific research into photosynthesis is vitally important. If we can understand and control the intricacies of the photosynthetic process, we can learn how to increase crop yields of food, fibre, wood, and fuel, and how to better use our lands. The energy-harvesting secrets of plants can be adapted to man-made systems, which provide new, efficient ways to collect and use solar energy. These same natural "technologies" can help point the way to the design of new, faster, and more compact computers, and even to new medical breakthroughs. Because photosynthesis helps control the makeup of our atmosphere, understanding photosynthesis is crucial to understanding how carbon dioxide and other "greenhouse gases" affect the global climate. In this document, we will briefly explore each of the areas mentioned above, and illustrate how photosynthesis research is critical to maintaining and improving our quality of life.

The ultimate source of energy for life on Earth is the sun. Plants are able to transform the light energy from the sun into chemical energy through a process called photosynthesis. Through absorption of light energy, plants, algae, and a few types of bacteria transform carbon dioxide and water into carbohydrates. Carbohydrates serve as fuel for their growth and metabolism. Other organisms benefit indirectly from photosynthesis. Oxygen is a waste product of photosynthesis. By releasing oxygen into the environment, plants make the air breathable for many other life forms.

The presence of oxygen in relation to plants was first demonstrated by the English chemist, Joseph Priestley in 1772. He noted that air that had been depleted through burning candles could be made breathable again by plants. Seven years later, a Dutch doctor, Jan Ingenhousz, showed that plants required sunlight in order to make air breathable. Ingenhousz also demonstrated that it was only the green parts of plants that had this ability. However, neither scientist knew anything of oxygen's existence. In 1782, Swiss scientist Jean Senebier proved that the plants also needed "fixed air"—what is now called carbon dioxide—in addition to sunlight. His discovery was followed in 1804 by another Swiss scientist, Nicholas Theodore de Saussure, proving that water was also a necessary factor.

By the early 19th century, scientists had at least a general understanding of photosynthesis. They knew that green plants, if given carbon dioxide, water, and sunlight, would produce organic material and oxygen. Building on this foundation, scientists began to tease apart the details of how the plants accomplished this process.

A big step forward was the discovery that chloroplasts were the site of oxygen generation in plant cells. Chloroplasts are cell organelles comparable to the mitochondria of animal cells. These organelles contain a cell's energy-generating mechanisms. Discovering the function of chloroplasts prompted scientists to scrutinise the organelle's structure and chemical makeup.

Not all cells capable of photosynthesis have chloroplasts, however. Cells are generally divided into two categories: prokaryotes and eukaryotes. Eukaryotes have chloroplasts, but prokaryotes do not. The difference between eukaryotes and prokaryotes is that eukaryotes have a nucleus, while prokaryotes do not. The nucleus is a cellular structure that stores genetic material separately from the rest of the cell interior. Plants and certain types of algae are eukaryotes. Blue-green algae and photosynthetic bacteria are prokaryotes. The following discussion of photosynthesis will concentrate on photosynthetic eukaryotes—i.e., plants and algae. Chloroplasts contain chlorophyll, a green-coloured pigment. Chlorophyll is the key molecule in photosynthesis. There are several types of chlorophyll, but the predominant form is chlorophyll a. Chlorophyll molecules are composed of a central porphyrin ring and side chains. These side chains differ slightly between types of chlorophyll, but the general appearance of the molecule is very similar. The porphyrin ring is actually a complex multi-ring structure composed of carbon, hydrogen, nitrogen, and oxygen atoms.

Light is broken down into units called photons, which move, in waves. The distance between waves is called the wavelength. Visible light contains wavelengths ranging from 350 nm to 800 nm. Longer wavelengths contain less energy; short wavelengths contain more. Wavelengths correspond with the colours of the rainbow—called the visible light spectrum—and each colour represents a different energy level.

The colours that an object appears to be are those that are reflected; colours that are absorbed are not seen. Chlorophyll absorbs energy from the red portion of the spectrum, approximately 680-700 nm, and reflects the green portion. For that reason, the leaves and stems of plants appear green. When a photon is absorbed by a chlorophyll molecule, its energy causes the chlorophyll to enter an energy-rich excited state.

Within a chloroplast, chlorophyll molecules are densely packed together. Because the chlorophyll molecules are so close to each other, the energy they absorb circulates from molecule to molecule. This energy is eventually transmitted to a reaction centre, thought to be a special arrangement of chlorophyll and protein molecules. There are two types of reaction centres in chloroplasts, each of which corresponds with a photosystem. The P700 reaction centre is associated with photosystem I; P680 is linked with photosystem II. The names of the photosystems arise from the wavelength of light, which they most efficiently absorb.

Both photosystems serve as trigger points for electron transport chains. Electron transport chains operate through the reduction and oxidation of chemicals or chemical complexes that make up its links. Although reduction usually means that something is lost, it has the opposite meaning in chemistry. As a chemical term, reduction means that an electron is gained. Oxidation means that an electron is lost.

When P680 absorbs the energy of a photon, it goes from

a ground state to an excited state. This change is seen at the atomic level. In the ground state, the electrons of an atom are at a low- energy configuration. In an excited state, an electron jumps from a normal low-energy position, or orbital, to a high-energy orbital. The high-energy position is unstable, and the electron is quickly transferred to another chemical, pheophytin. Pheophytin is the first link in the photosystem II electron transport chain.

As P680 returns to its ground state, it picks up an electron to replace the one transferred to pheophytin. This electron is taken from a water molecule. In chemical terms, P680 is reduced and water is oxidised. The water molecule is split into hydrogen ions (positively charged atoms) and oxygen. For every two water molecules broken down in this manner, four hydrogen ions and one oxygen molecule are produced. The hydrogen ions can be used for other reactions, but the molecular oxygen is released as a useless waste product.

The electron transport chain that picks up an electron from P680, leads to P700. The chain itself comprises several chemical links. The extra electron is passed from one link to the next in a series of reductions and oxidations. As the electron proceeds to P700, it encounters a chemical complex called cytochrome bf. The final link in the photosystem II electron transport chain is plastocyanin.

The P700 reacts to a photon in the same manner as P680. When the photon's energy is absorbed, P700 is transformed from its ground state to an excited state. In its excited state, P700 transfers an electron to the photosystem I electron transport chain. In returning to its ground state, P700 is reduced by plastocyanin, courtesy of photosystem II.

The photosystem I electron transport chain leads to an enzyme called ferredoxin-NADP reductase. (Enzymes are protein molecules that trigger speedy chemical reactions.) Once this enzyme has the extra electron available, it is able to transform $NADP^+$ to NADPH and a hydrogen ion. NADPH is the abbreviation for nicotinamide adenine dinucleotide phosphate. This molecule is a high-energy fuel used for certain chemical reactions in the cell.

Under certain circumstances, P700 is raised to its excited state and the electron passes through only the first few links of the photosystem I electron transport chain. Instead of proceeding to the last link, the electron is shunted over to the cytochrome bf complex of photosystem II. The end result of this process—called cyclic photophosphorylation—is the production of a different high-energy fuel. This fuel is adenosine triphosphate, or ATP. Both ATP and NADPH are use to drive the chemical reactions that produce carbohydrates.

Both photosystem I and photosystem II require sunlight. As a result, this part of photosynthesis falls under the heading of light reactions. The actual generation of carbohydrates is a dark reaction, because it can occur in the absence of light. Carbohydrate construction, or synthesis, is accomplished through the Calvin cycle. The Calvin cycle is a series of chemical reactions through which carbon dioxide is used to build glucose. Glucose, in turn, is used as an eventual building block for sucrose, starch, and other carbohydrates.

The first step of the Calvin cycle is the addition of carbon dioxide to an acceptor molecule. This step was first described by Melvin Calvin in the 1940s, and the entire cycle is named in his honour. The cycle begins with the addition of carbon dioxide to another molecule, ribulose-1,5-bisphosphate. This reaction is triggered, or catalysed, the enzyme ribulose 1,5-bisphosphate carboxylase/oxygenase—often called by its short name, rubisco. Rubisco is able to add either carbon dioxide or molecular oxygen to a ribulose-1,5-bisphosphate molecule. The addition of carbon dioxide leads into the Calvin cycle; the addition of oxygen leads to a different process called photorespiration.

When rubisco binds carbon dioxide to ribulose-1, 5-bisphosphate, a six-carbon molecule is temporarily formed. This molecule quickly splits into two molecules of 3-phosphoglycerate. At this stage, ATP enters the cycle by transferring a phosphate group to the molecule. The resulting molecule is reduced by NADPH to form glyceraldehyde-3-phosphate. At this point, glyceraldehyde-3-phosphate can go in one of two directions.

A certain quantity of the molecule is dedicated to completing the cycle by regenerating the ribulose-1,5-bisphosphate stores. The remaining glyceraldehyde-3-phosphate continues on towards glucose, sucrose, starch, and other carbohydrate production.

Some plants have the ability to concentrate carbon dioxide levels within their cells before it enters the Calvin cycle. This ability helps circumvent photorespiration. As mentioned above, photorespiration occurs when rubisco adds oxygen, rather than carbon dioxide, to ribulose-1,5- bisphosphate. This is a wasteful reaction for the plant because it uses up ATP and reduces carbohydrate synthesis.

Generally rubisco will use carbon dioxide in preference to oxygen, but in hot, sunny climates, the carbon dioxide near plants is rapidly used up. With lower levels of carbon dioxide in the area, oxygen uptake would increase and photorespiration would occur.

Some plants have evolved a mechanism to reduce photorespiration. These plants are called C_4 plants, because they create a four-carbon molecule as an intermediary between carbon dioxide and the start of the Calvin cycle. The C_4 plants concentrate carbon dioxide levels prior to the Calvin cycle through the Hatch-Slack pathway. The Hatch-Slack pathway is named for its discovers, Australian biochemists, Marshall Hatch and Roger Slack. Examples of C_4 plants are corn, sorghum, and sugar cane, both of which do well in hot, sunny conditions. Plants which add carbon dioxide directly to the Calvin cycle are called C_3 plants, because they create the three-carbon molecule, 3-phosphoglycerate. The C_3 plants do best in temperate climates. Wheat is an example of a C_3 plant. Regardless of how carbon dioxide enters the Calvin cycle, the products are the same.

Photosynthesis and Food

All of our biological energy needs are met by the plant kingdom, either directly or through herbivorous animals. Plants in turn obtain the energy to synthesise foodstuffs via photosynthesis. Although plants draw necessary materials from the soil and water and carbon dioxide from the air, the energy needs of the plant are filled by sunlight. Sunlight is pure energy. However, sunlight itself is not a very useful form of energy; it cannot be eaten, it cannot turn dynamos, and it cannot be stored. To be beneficial, the energy in sunlight must be converted to other forms. This is what photosynthesis is all about. It is the process by which plants change the energy in sunlight to kinds of energy that can be stored for later use. Plants carry out this process in photosynthetic reaction centres. These tiny units are found in leaves, and convert light energy to chemical energy, which is the form used by all living organisms. One of the major energy-harvesting processes in plants involves using the energy of sunlight to convert carbon dioxide from the air into sugars, starches, and other high-energy carbohydrates. Oxygen is released in the process. Later, when the plant needs food, it draws upon the energy stored in these carbohydrates. We do the same. When we eat a plate of spaghetti, our bodies oxidise or "burn" the starch by allowing it to combine with oxygen from the air. This produces carbon dioxide, which we exhale, and the energy we need to survive. Thus, if there is no photosynthesis, there is no food. Indeed, one widely accepted theory explaining the extinction of the dinosaurs suggests that a comet, meteor, or volcano ejected so much material into the atmosphere that the amount of sunlight reaching the earth was severely reduced. This in turn caused the death of many plants and the creatures that depended upon them for energy.

Photosynthesis and Energy

One of the carbohydrates resulting from photosynthesis is cellulose, which makes up the bulk of dry wood and other plant material. When we burn wood, we convert the cellulose back to carbon dioxide and release the stored energy as heat. Burning fuel is basically the same oxidation process that occurs in our bodies; it liberates the energy of "stored sunlight" in a useful form, and returns carbon dioxide to the atmosphere. Energy from burning "biomass" is important in many parts of the world.

In developing countries, firewood continues to be critical to survival. Ethanol (grain alcohol) produced from sugars and starches by fermentation is a major automobile fuel in Brazil, and is added to gasoline in some parts of the United States to help reduce emissions of harmful pollutants. Ethanol is also readily converted to ethylene, which serves as a feedstock to a large part of the petrochemical industry. It is possible to convert cellulose to sugar, and then into ethanol; various microorganisms carry out this process. It could be commercially important one day.

Our major sources of energy, of course, are coal, oil and natural gas. These materials are all derived from ancient plants and animals, and the energy stored within them is chemical energy that originally came from sunlight through photosynthesis. Thus, most of the energy we use today is originally solar energy!

Photosynthesis, Fibre and Materials

Wood, of course, is not only burned, but is an important material for building and many other purposes. Paper, for example, is nearly pure photosynthetically produced cellulose, as is cotton and many other natural fibres. Even wool production depends on photosynthetically-derived energy. In fact, all plant and animal products including many medicines and drugs require energy to produce, and that energy comes ultimately from sunlight via photosynthesis. Many of our other materials needs are filled by plastics and synthetic fibres which are produced from petroleum, and are thus also photosynthetic in origin. Even much of our metal refining depends ultimately on coal or other photosynthetic products. Indeed, it is difficult to name an economically important material or substance whose existence and usefulness is not in some way tied to photosynthesis.

Photosynthesis and Environment

Currently, there is a lot of discussion concerning the possible effects of carbon dioxide and other "greenhouse gases" on the environment. Photosynthesis converts carbon dioxide from the air to carbohydrates and other kinds of "fixed" carbon and releases oxygen to the atmosphere. When we burn firewood, ethanol, or coal, oil and other fossil fuels, oxygen is consumed, and carbon dioxide is released back to the atmosphere. Thus, carbon dioxide which was removed from the atmosphere over millions of years is being replaced very quickly through our consumption of these fuels. The increase in carbon dioxide and related gases is bound to affect our atmosphere. Will this change be large or small, and will it be harmful or beneficial? These questions are being actively studied by many scientists today. The answers will depend strongly on the effect of photosynthesis carried out by land and sea organisms. As photosynthesis consumes carbon dioxide and releases oxygen, it helps counteract the effect of combustion of fossil fuels. The burning of fossil fuels releases not only carbon dioxide, but also hydrocarbons, nitrogen oxides, and other trace materials that pollute the atmosphere and contribute to long-term health and environmental problems. These problems are a consequence of the fact that nature has chosen to implement photosynthesis through conversion of carbon dioxide to energy-rich materials such as carbohydrates. Can the principles of photosynthetic solar energy harvesting be used in some way to produce non-polluting fuels or energy sources? The answer, as we shall see, is yes.

Why to Study Photosynthesis?

Because our quality of life, and indeed our very existence, depends on photosynthesis, it is essential that we understand it. Through understanding, we can avoid adversely affecting the process and precipitating environmental or ecological disasters. Through understanding, we can also learn to control photosynthesis, and thus enhance production of food, fibre and energy. Understanding the natural process, which has been developed by plants over several billion years, will also allow us to use the basic chemistry and physics of photosynthesis for other purposes, such as solar energy conversion, the design of electronic circuits, and the development of medicines and drugs.

Photosynthesis and Agriculture

Although photosynthesis has interested mankind for eons, rapid progress in understanding the process has come in the last few years. One of the things we have learned is that overall, photosynthesis is relatively inefficient. For example, based on the amount of carbon fixed by a field of corn during a typical growing season, only about 1 - 2 per cent of the solar energy falling on the field is recovered as new photosynthetic products. The efficiency of uncultivated plant life is only about 0.2 per cent. In sugar cane, which is one of the most efficient plants, about 8 per cent of the light absorbed by the plant is preserved as chemical energy. Many plants, especially those that originate in the temperate zones such as most of the United States, undergo a process called photorespiration. This is a kind of "short circuit" of photosynthesis that wastes much of the plants' photosynthetic energy. The phenomenon of photorespiration including its function, if any, is only one of many riddles facing the photosynthesis researcher.

If we can fully understand processes like photorespiration, we will have the ability to alter them. Thus, more efficient plants can be designed. Although new varieties of plants have been developed for centuries through selective breeding, the techniques of modern molecular biology have speeded up the process tremendously. Photosynthesis research can show us how to produce new crop strains that will make much better use of the sunlight they absorb. Research along these lines is critical, as recent studies show that agricultural production is levelling off at a time when demand for food and other agricultural products is increasing rapidly.

Because plants depend upon photosynthesis for their survival, interfering with photosynthesis can kill the plant. This is the basis of several important herbicides, which act by preventing certain important steps of photosynthesis. Understanding the details of photosynthesis can lead to the design of new, extremely selective herbicides and plant growth regulators that have the potential of being environmentally safe (especially to animal life, which does not carry out photosynthesis). Indeed, it is possible to develop new crop plants that are immune to specific herbicides, and to thus achieve weed control specific to one crop species.

Photosynthesis and Energy Production

Most of our current energy needs are met by photosynthesis. Increasing the efficiency of natural photosynthesis can also increase production of ethanol and other fuels derived from agriculture. However, knowledge gained from photosynthesis research can also be used to enhance energy production in a much more direct way. Although the overall photosynthesis process is relatively wasteful, the early steps in the conversion of sunlight to chemical energy are quite efficient.

Why not learn to understand the basic chemistry and physics of photosynthesis, and use these same principles to build man-made solar energy harvesting devices? This has been a dream of chemists for years, but is now close to becoming a reality. In the laboratory, scientists can now synthesise artificial photosynthetic reaction centres which rival the natural ones in terms of the amount of sunlight stored as chemical or electrical energy. More research will lead to the development of new, efficient solar energy harvesting technologies based on the natural process.

Role of Photosynthesis in Control of Environment

How does photosynthesis in temperate and tropical forests and in the sea affect the quantity of greenhouse gases in the atmosphere? This is an important and controversial issue today. Photosynthesis by plants removes carbon dioxide from the atmosphere and replaces it with oxygen. Thus, it would tend to ameliorate the effects of carbon dioxide released by the burning of fossil fuels. However, the question is complicated by the fact that plants themselves react to the amount of carbon dioxide in the atmosphere.

Some plants, appear to grow more rapidly in an atmosphere rich in carbon dioxide, but this may not be true of all species. Understanding the effect of greenhouse gases requires a much better knowledge of the interaction of the plant kingdom with carbon dioxide than we have today. Burning plants and plant products such as petroleum releases carbon dioxide and other by-products such as hydrocarbons and nitrogen oxides. However, the pollution caused by such materials is not a necessary product of solar energy utilisation. The artificial photosynthetic reaction centres produce energy without releasing any by-products other than heat. They hold the promise of producing clean energy in the form of electricity or hydrogen fuel without pollution. Implementation of such solar energy harvesting devices would prevent pollution at the source, which is certainly the most efficient approach to control.

Photosynthesis and Electronics

At first glance, photosynthesis would seem to have no association with the design of computers and other electronic devices. However, there is potentially a very strong connection. A goal of modern electronics research is to make transistors and other circuit components as small as possible. Small devices and short connections between them make computers faster and more compact.

The smallest possible unit of a material is a molecule (made up of atoms of various types). Thus, the smallest conceivable transistor is a single molecule (or atom). Many researchers today are investigating the intriguing possibility of making electronic components from single molecules or small groups of molecules.

Another very active area of research is computers that use light, rather than electrons, as the medium for carrying information. In principle, light-based computers have several advantages over traditional designs, and indeed many of our telephone transmission and switching networks already operate through fibre optics. What does this have to do with photosynthesis? It turns out that photosynthetic reaction centres are natural photochemical switches of molecular dimensions.

Learning how plants absorb light, control the movement of the resulting energy to reaction centres, and convert the light energy to electrical, and finally chemical energy can help us understand how to make molecular-scale computers. In fact, several molecular electronic logic elements based on artificial photosynthetic reaction centres have already been reported in the scientific literature.

Photosynthesis and Medicines

Light has a very high energy content, and when it is absorbed by a substance this energy is converted to other forms. When the energy ends up in the wrong place, it can cause serious damage to living organisms. Aging of the skin and skin cancer are only two of many deleterious effects of light on humans and animals.

Because plants and other photosynthetic species have been dealing with light for eons, they have had to develop photoprotective mechanisms to limit light damage. Learning about the causes of light- induced tissue damage and the details of the natural photoprotective mechanisms can help us can find ways to adapt these processes for the benefit of humanity in areas far removed from photosynthesis itself. For example, the mechanism by which sunlight absorbed by photosynthetic chlorophyll causes tissue damage in plants has been harnessed for medical purposes. Substances related to chlorophyll localise naturally in cancerous tumour tissue.

Illumination of the tumors with light then leads to photochemical damage which can kill the tumour while leaving surrounding tissue unharmed. Another medical application involves using similar chlorophyll relatives to localise in tumour tissue, and thus act as dyes which clearly delineate the boundary between cancerous and healthy tissue. This diagnostic aid does not cause photochemical damage to normal tissue because the principles of photosynthesis have been used to endow it with protective agents that harmlessly convert the absorbed light to heat. The above examples illustrate the importance of photosynthesis as a natural process and the impact that it has on all of our lives. Research into the nature of photosynthesis is crucial because only by understanding photosynthesis can we control it, and harness its principles for the betterment of mankind. Science has only recently developed the basic tools and techniques needed to investigate the intricate details of photosynthesis. It is now time to apply these tools and techniques to the problem, and to begin to reap the benefits of this research.

Basic Needs for Photosynthesis

Plants, as well as some Protists and Monerans, can take small molecules from the environment and bind them together using the energy of light. The incoming light energy is transformed into the energy holding the new molecules together, and the organisms use those molecules as an energy "fuel." The basic process can be represented this way:

$$CO_2 + H_2O \xrightarrow{\text{light}} C_6H_{12}O_6 + O_2$$

Carbon Dioxide, Water → Glucose (sugar), Oxygen

In the case of organisms that live in water, the carbon dioxide and water are from their immediate surroundings; for most land plants, the water is absorbed from the soil and the carbon dioxide from the atmosphere.

The glucose is used for two major purposes:

(1) It serves as an energy reserve for periods of darkness (don't forget that photosynthesisers, like any living things, require energy and get it through respiration processes, commonly aerobic respiration; and

(2) It is used as a major component of structure: the cell walls that surround almost all photosynthetic cells are made of starches, huge molecules made up of hundreds, commonly thousands, of sugar molecules bound together. This is why plant fibres are great sources of nutrition if you can break them down. Breaking down plant fibres is chemically difficult - we humans can't, being limited to the more digestible starches put into seeds and fruits and tubers. Plants use these starches themselves as sources of sugar fuels, and so build them into a molecule that is much easier to break down

Keep in mind that photosynthetic organisms are still living things, with protein-based chemistry, which means that they have nutritional requirements beyond carbon dioxide and water. Proteins, unlike sugars and starches, contain a significant amount of *nitrogen*, which usually needs to be absorbed as *nitrates* (a nitrogen-oxygen molecule) to be usable. The production and use of glucose for energy also requires *ATP* as an energy carrier; ATP contains *phosphorus*, usually absorbed as *phosphates* (a phosphorus-oxygen molecule). Anyone who takes care of plants knows that nitrates and phosphates are important ingredients in fertilizers. Most photosynthesisers have a few critical molecules that contain other materials as well, such as iron, or need small ions, such as sodium, for some of their chemical processes.

Light as an Energy Source

Light can be understood as a combination of energy waves travelling outward from a light source, or as small packets moving from that source at the speed of light (each wave peak would correspond to a single packet). Light always travels at the speed of light, altering only for the material through which it's moving (it goes slower in water, for instance), so a segment of a light beam with wave peaks more separated (a longer *wavelength*) would have fewer peaks absorbed by a surface (a lower *frequency*) in any given amount of time, and would hit that surface with less energy.

This means *short wavelength = high frequency = more energy, long wavelength = low frequency = less energy.* The only reason that this is important is that sunlight contains a fairly wide range of energy frequencies, but only a few are absorbed and used by *chlorophyll,* the energy-capturing molecule of photosynthesis.

You can tell a few of the frequencies that are not absorbed by chlorophyll (and a few other light-absorbing molecules) by looking at a plant. That green you see is part of the *reflected frequencies* of light. For the most part, absorption of light is used in an energy conversion that "spits" electrons through a system from the "excited" chlorophyll molecules.

Although land plants absorb a variety of light frequencies, all frequencies are not equally powerful or useful: while plants can absorb both red frequencies and purple frequencies, the purple have shorter wavelengths and carry more energy. This is one of the reasons why "plant lights" are distinctly purple.

It is not unusual for land plants to use molecular supplements to absorb some frequencies that chlorophyll can't, and feed more energy into the photosynthesis process; these *pigments* are commonly types of *carotenoids.* The colours of leaves in the autumn reveals the carotenoids that have always been there but have been covered by huge amount of chlorophyll.

Carotenoids can serve multiple roles: they can be photosynthetic aids, but they may also minimise light damage (animals use pigments, like the human tan-producing molecule melanin, for similar protection) or even function in fighting disease. Land plants may concentrate pigments, including carotenoids, in structure that need to stand out, such as the colours of flowers or mature fruits. These colours signal animals that food is available, and then the animals are used to carry pollen or seeds.

Photosynthesis is a Two-step Process

This will be a very "bare bones" summation of the photosynthetic chemistry: Photosynthesis breaks down into a *Light-Dependent Reaction* and a *Light-Independent Reaction.* The light-dependent reaction uses, not too surprisingly, light, but it uses the water (actually, the hydrogen part of the water, which releases the oxygen). This part of photosynthesis shifts the light energy into energy of several carriers, including a lot of ATP. These energy carriers drive the light-independent reaction, which uses the carbon dioxide and actually makes the glucose. For most plants on a typically sunny day, when the sun goes down the light-dependent reaction stops, but the backlog of energy carriers it has made may keep the light-independent reaction going until the middle of the night.

It's interesting that photosynthesis is a sort of "mirror image" to *aerobic respiration:*

$$C_6H_{12}O_6 + O_2 \text{ energy to ATP} \rightarrow CO_2 + H_2O$$

It uses glucose and oxygen and accesses the energy stored in the glucose, releasing carbon dioxide and water. Even in some of the more detailed steps, the two processes still "reflect" each other.

Algae

Plants, like all living things, are descended from ancient organisms that evolved and developed in the oceans. The ocean-living plants that represent land plants ancestors are simple, so much so that they are often classified as Protistans-little multicellular complexity was needed to float in the light.

Even in the ocean, though, there are a number of different niches available for plants—in the shallows, many algae varieties exist that can attach to submerged objects, probably a feature that helped the land plants' direct ancestors stick out of the water into the air. But niches also in the deeper water, where there is a connection to the nature of how light interacts with water.

The surface of the water reflects some light, but the frequencies that get through get absorbed—you already know that the deeper you go, the darker it gets. What you may not know is that not all frequencies get absorbed equally—the frequencies that green plants use don't get very far beneath the surface, but frequencies in the bluish and greenish ranges penetrate much further. This has spurred the evolution of several classes of various coloured algae with different chlorophylls, capable of using these deeper-reaching frequencies to grow where the green algae can't. *Red algaes* especially can occupy greater depths than the others, although *Golden algaes* can occupy a bit of a middle zone between green and red. The colour of *Brown algaes* seems more to do with other pigments and is apparently not really depth-related.

Chlorophyll Biosynthesis

All living organisms contain tetrapyrroles. In plants, chlorophyll (chlorophyll a plus chlorophyll b) is the most abundant and probably most important tetrapyrrole. It is involved in light absorption and energy transduction during photosynthesis. Chlorophyll is synthesised from the intact carbon skeleton of glutamate via the C_5 pathway. This pathway takes place in the chloroplast. It is the aim of this review to summarise the current knowledge on the biochemistry and molecular biology of the C_5-pathway enzymes, their regulated expression in response to light, and the impact of chlorophyll biosynthesis on chloroplast development. Particular emphasis will be placed on the key regulatory steps of chlorophyll biosynthesis in higher plants, such as 5-aminolevulinic acid formation, the production of Mg(2+)- protoporphyrin IX, and light-dependent protochlorophyllide reduction.

Thioredoxins are small multifunctional redox active proteins widely if not universally distributed among living organisms. In chloroplasts, two types of thioredoxins (*f* and *m*) coexist and play central roles in regulating enzyme activity. Reduction of thioredoxins in chloroplasts is catalysed by an iron-sulphur disulphide enzyme, ferredoxin-thioredoxin reductase, that receives photosynthetic electrons from ferredoxin, thereby providing a link between light and enzyme activity. Chloroplast thioredoxins function in the regulation of the Calvin cycle and associated processes. However, the relatively small number of known thioredoxin-linked proteins (about 16) raised the possibility that others remain to be identified. To pursue this opportunity, we have mutated thioredoxins *f* and *m*, such that the buried cysteine of the active disulphide has been replaced by serine or alanine, and bound them to affinity columns to trap target proteins of chloroplast stroma. The covalently linked proteins were eluted with DTT, separated on gels, and identified by mass spectrometry.

This approach led to the identification of 15 potential targets that function in 10 chloroplast processes not known to be thioredoxin linked. Included are proteins that seem to function in plastid-to-nucleus signalling and in a previously unrecognised type of oxidative regulation. Approximately two-third of these targets contained conserved cysteines. We also identified 11 previously unknown and 9 confirmed target proteins that are members of pathways known to be regulated by thioredoxin. In contrast to results with individual enzyme assays, specificity for thioredoxin f or m was not observed on affinity chromatography.

Evolution of Photosynthesis

Historical Backdrop

In the 1770s Joseph Priestley, an English chemist and clergyman, performed experiments showing that plants release a type of air that allows combustion. He demonstrated this by burning a candle in a closed vessel until the flame went out. He placed a sprig of mint in the chamber and after several days showed that the candle could burn again. Although Priestley did not know about molecular oxygen, his work showed that plants release oxygen into the atmosphere. It is noteworthy that over 200 years later, investigating the mechanism by which plants produce oxygen, is one of the most active areas of photosynthetic research. Building on the work of Priestley, Jan Ingenhousz, a Dutch physician, demonstrated that sunlight was necessary for photosynthesis and that only the green parts of plants could release oxygen. During this period Jean Senebier, a Swiss botanist and naturalist, discovered that CO_2 is required for photosynthetic growth and Nicolas- Theodore de Saussure, a Swiss chemist and plant physiologist, showed that water is required. It was not until 1845 that Julius Robert von Mayer, a German physician and physicist, proposed that photosynthetic organisms convert light energy into chemical free energy. An interesting time line of the history of photosynthesis has been presented by Huzisige and Ke (1993).

By the middle of the nineteenth century the key features of plant photosynthesis were known, namely, that plants could use light energy to make carbohydrates from CO_2 and water. The empirical equation representing the net reaction of photosynthesis for oxygen evolving organisms is:

$$CO_2 + 2H_2O + \text{Light Energy} \rightarrow CH_2O + O_2 + H_2O$$

where CH_2O represents a carbohydrate (e.g., glucose, a six-carbon sugar). The synthesis of carbohydrate from carbon and water requires a large input of light energy. The standard free energy for the reduction of one mole of CO2 to the level of glucose is +478 kJ/mol. Because glucose, a six carbon sugar, is often an intermediate product of photosynthesis, the net equation of photosynthesis is frequently written as:

$$6CO_2 + 12H_2O + \text{Light Energy} \rightarrow C6H1_2O_6 + 6O_2 + 6H_2O$$

The standard free energy for the synthesis of glucose is +2,870 kJ/mol.

Not surprisingly, early scientists studying photosynthesis concluded that the O_2 released by plants came from CO_2, which was thought to be split by light energy. In the 1930s comparison of bacterial and plant photosynthesis lead Cornelis van Niel to propose the general equation of photosynthesis that applies to plants, algae and photosynthetic bacteria (discussed by Wraight, 1982). Van Niel was aware that some photosynthetic bacteria could use hydrogen sulphide (H_2S) instead of water for photosynthesis and that these organisms released sulphur instead of oxygen. Van Niel, among others, concluded that photosynthesis depends on electron donation and acceptor reactions and that the O_2 released during photosynthesis comes from the oxidation of water. Van Niel's generalised equation is:

$$CO_2 + 2H_2A + \text{Light Energy} \rightarrow CH_2O + 2A + H_2O$$

In oxygenic photosynthesis, 2A is O_2, whereas in anoxygenic photosynthesis, which occurs in some photosynthetic bacteria, the electron donor can be an inorganic hydrogen donor, such as H_2S (in which case A is elemental sulphur) or an organic hydrogen donor such as succinate (in which case, A is fumarate). Experimental evidence that molecular oxygen came from water was provided by Hill and Scarisbrick (1940) who demonstrated oxygen evolution in the absence of CO_2 in illuminated chloroplasts and by Ruben *et al.* (1941) who used 18O enriched water.

The biochemical conversion of CO_2 to carbohydrate is a reduction reaction that involves the rearrangement of covalent bonds between carbon, hydrogen and oxygen. The energy for the reduction of carbon is provided by energy rich molecules that are produced by the light driven electron transfer reactions. Carbon reduction can occur in the dark and involves a series of biochemical reactions that were elucidated by Melvin Calvin, Andrew Benson and James Bassham in the late 1940s and 1950s. Using the radioisotope 14C, most of the intermediate steps that result in the production of carbohydrate were identified. Calvin was awarded the Nobel Prize for Chemistry in 1961 for this work.

In 1954 Daniel Arnon and co-workers discovered that plants, and A. Frenkel discovered that photosynthetic bacteria, use light energy to produce ATP, an organic molecule that serves as an energy source for many biochemical reactions. During the same period L.N.M. Duysens showed that the primary photochemical reaction of photosynthesis is an oxidation/reduction reaction that occurs in a protein complex (the reaction centre). Over the next few years the work of several groups, including those of Robert Emerson, Bessel Kok, L.N.M. Duysens, Robert Hill and Horst Witt, combined to prove that plants, algae and cyanobacteria require two reaction centres, photosystem II and photosystem I, operating in series (Duysens, 1989; Witt, 1991).

In 1961 Peter Mitchell suggested that cells can store energy by creating an electric field or a proton gradient across a membrane. Mitchell's proposal that energy is stored as an electrochemical gradient across a vesicular membrane opened the door for understanding energy transformation by membrane systems. He was awarded the Nobel Prize in Chemistry in 1978 for his theory of chemiosmotic energy transduction (Mitchell, 1961).

Most of the proteins required for the conversion of light energy and electron transfer reactions of photosynthesis are located in membranes. Despite decades of work, efforts to determine the structure of membrane bound proteins had little success. This changed in the 1980s when Johann Deisenhofer, Hartmut Michel, Robert Huber and co-workers determined the structure of the reaction centre of the purple bacterium Rhodospeudomonas viridis. (Deisenhofer *et al.*, 1984, 1985; Deisenhofer and Michel, 1993). They were awarded the Nobel Prize for Chemistry in 1988 for their work, which has provided insight into the relationship between structure and function in membrane-bound proteins.

A key element in photosynthetic energy conversion is electron transfer within and between protein complexes and simple organic molecules. The electron transfer reactions are rapid (as fast as a few picoseconds) and highly specific. Much of our current understanding of the physical principles that guide electron transfer is based on the pioneering work of Rudolph A. Marcus (Marcus and Sutin, 1985), who received the Nobel Prize in Chemistry in 1992 for his contributions to the theory of electron transfer reaction in chemical systems.

On the basis of his studies with photosynthetic bacteria,

Van Niel proposed that the oxygen which plants produce during photosynthesis is derived from water, not from carbon dioxide. In the following years, this hypothesis has proven true. Van Niel's brilliant insight was a major contribution to our modern understanding of photosynthesis.

The study of photosynthesis is currently a very active area of research in biology. Hartmut Michel and Johann Deisenhofer recently made a very important contribution to our understanding of photosynthesis. They made crystals of the photosynthetic reaction centre from *Rhodopseudomonas viridis*, an anaerobic photosynthetic bacterium, and then used x-ray crystallography to determine its three-dimensional structure. In 1988, they shared the Nobel Prize in Chemistry with Robert Huber for this ground-breaking research.

Modern plant physiologists commonly think of photosynthesis as consisting of two separate series of interconnected biochemical reactions, the light reactions and the dark reactions. The light reactions use the light energy absorbed by chlorophyll to synthesise labile high energy molecules. The dark reactions use these labile high energy molecules to synthesise carbohydrates, a stable form of chemical energy which can be stored by plants. Although the dark reactions do not require light, they often occur in the light because they are dependent upon the light reactions. In higher plants and algae, the light and dark reactions of photosynthesis occur in chloroplasts, specialised chlorophyll-containing intracellular structures which are enclosed by double membranes.

Research on photosynthesis has been closely linked to knowledge of the growth cycles and physical structure of plants. In the 1640s, the work of both Johannes (Jan) Baptista van Helmont (1577-1644) and English clergyman and physiologist (a person specialising the study of the processes of living things, noted by Bruno (2001)) Stephen Hales indicated that plants require air and water to grow. In the 1700s, chemists began to identify the individual gases involved in the processes of combustion, respiration, and photosynthesis. Joseph Priestley (1733 - 1804) demonstrated that green plants can replenish stale, or oxygen-poor, air so that it is capable of supporting combustion and respiration.

Dutch doctor and plant physiologist Jan Ingenhousz (1730 - 1799), inspired by Priestley's research, later learned that only the green parts of plants can revitalise stale air — that is, take in carbon dioxide and release oxygen — and that they do so only in the presence of sunlight. This was the first indication of light's role in the photosynthetic process. Ingenhousz also discovered that only the light of the Sun — and not the heat it generates — is necessary for photosynthesis.

In the nineteenth century, research on photosynthesis centred on the chemical processes in which carbon is "fixed" in carbohydrates. In the late 1800s, German botanist (a person specialising the study of plants, noted by Bruno (2001)) Julius Von Sachs (1832 - 1897) suggested that starch is a product of carbon dioxide. He also argued in 1865, that, in the presence of light, chlorophyll catalyses photosynthetic reactions, and he discovered the chlorophyll containing chloroplasts. In the 1880s, German physiologist Theodor Wilhelm Engelmann (1843 - 1909) showed that the light reactions, which capture solar energy and convert it into chemical energy, occur within the chloroplasts and respond only to the red and blue hues of natural light.

It was not until the twentieth century that scientists began to understand the complex biochemistry of photosynthesis. Richard Willstatter recognised that there were two major types of chlorophyll in land plants: blue-green, or "a" type, and yellow-green, or "b" type. Martin David Kamen, a Canadian-born American biochemist, used the isotope (forms of an element having the same atomic number but a different atomic weight due to a different number of neutrons (Mader, 1990)) oxygen-18 to trace the chemical's role in the process. He confirmed that the oxygen created during photosynthesis comes only from the water molecules.

Germ biochemist Otto Warburg found that, under suitable conditions, the efficiency of the photosynthetic process can approach 100 per cent, meaning that nearly all of the sun's energy is converted to chemical energy.

In 1940, the discovery of carbon-14, a radioactive isotope of carbon isolated by Kamen, allowed for more detailed studies of photosynthesis. Using carbon-14, Melvin Calvin was able to trace carbon's path through the entire photosynthetic process. During the 1950s and 1960s, he confirmed that the light reactions involving chlorophyll instantly capture the sun's energy. Then he studied the subsequent dark reactions, so-called because they can take place without sunlight, find that carbohydrate molecules begin to form at this stage of the process. Working with green algal cells, Calvin interrupted the photosynthetic process at different stages and plunged the cells into an alcohol solution. Then, using the laboratory technique called paper chromatography, he analysed the cells and the chemicals that had been produced, identifying at least ten intermediate products that had been created within a few seconds. This series of reactions is now called the Calvin Benson Cycle.

In 1998, scientists at Arizona State University announced that they had created an artificial photosynthetic energy system. The cell-like machine used light to power the synthesis of ATP, a carrier of chemical energy in all organisms. The new technology could eventually lead to biological computers and new drugs.

Process of Evolution

The ability to convert light energy to chemical energy confers a significant evolutionary advantage to living organisms. Early photosynthetic systems, such as those from green and purple sulphur and green and purple non-sulphur bacteria, are thought to have been anoxygenic, using various molecules as electron donors. Green and purple sulphur bacteria are thought to have used hydrogen and sulphur as an electron donor. Green non-sulphur bacteria used various amino and other organic acids. Purple non-sulphur bacteria used a variety of non-specific organic molecules. The use of these molecules is consistent with the geological evidence that the atmosphere was highly reduced at that time.

Fossils have been found of what are thought to be filamentous photosynthetic organisms dating from 3.4 billion years ago (*New Scientist*, 19 Aug., 2006).

The oxygen in the atmosphere today exists due to the evolution of oxygenic photosynthesis, sometimes referred to as the oxygen catastrophe. Geological evidence suggests that oxygenic photosynthesis, such as that in cyanobacteria, became important during the Paleoproterozoic era around 2 billion years ago. Modern photosynthesis in plants and most photosynthetic prokaryotes is oxygenic. Oxygenic photosynthesis uses water as an electron donor which is oxidised into molecular oxygen by the absorption of a photon by the photosynthetic reaction centre.

Life began very early in Earth's history, perhaps before 3800 million years ago (Ma), and achieved remarkable levels of metabolic sophistication before the end of the Archean around 2500 (Ma). The great antiquity of our biosphere might indeed illustrate how easily life can arise on a habitable planet, but it also portends the challenges that confront our efforts to become intimately familiar with our earliest ancestors. The earliest sedimentary rocks have typically undergone extensive alteration by metamorphosis, taking a serious toll on microfossils. Fortunately, memories of our distant forebears are recorded not only in ancient rocks, but also in biological macromolecules and pathways. The two records are highly complementary: The geologic record offers the absolute timing of evolutionary innovations and their environmental context, while the living biochemical record can reveal the sequence of development of key pathways and biomolecules.

Xiong *et al.* have tapped the biological record to study the evolution of photosynthesis. They have obtained new sequence information for genes involved in photosynthesis and performed phylogenetic analyses on the major groups of photosynthetic bacteria. The study better defines the molecular origins of these groups and clarifies the great antiquity of anoxygenic photosynthesis.

When our biosphere developed photosynthesis, it developed an energy resource orders of magnitude larger than that available from oxidation-reduction reactions associated with weathering and hydrothermal activity. The significance of this innovation can be illustrated quantitatively for modern Earth. Hydrothermal sources deliver (0.13 to 1.1) x 1012 mol year-1 globally of reduced S, Fe_2+, Mn_2+, H_2, and CH_4; this is estimated to sustain at most about (0.2 to 2.0) x 1012 mol C year-1 of organic carbon production by microorganisms capable of using hydrothermal energy as their energy source.

In contrast, global photosynthetic productivity is estimated at 9000 x 1012 mol C year-1. Global thermal fluxes were greater in the distant geologic past, but the onset of oxygenic photosynthesis most probably increased global organic productivity by at least two to three orders of magnitude. This enormous productivity resulted principally from the ability of oxygenic photosynthetic bacteria to capture hydrogen for organic biosynthesis by cleaving water. This virtually unlimited supply of hydrogen freed life from its sole dependence upon abiotic chemical sources of reducing power, such as hydrothermal sources and weathering. Communities sustained by oxygenic photosynthesis could thrive wherever supplies of sunlight, moisture, and nutrients were sufficient.

Photosynthetic microbial communities have left a relatively robust fossil record, in part because their productivity was particularly high on stable submerged continental platforms and margins, and thus contributed to sediments with excellent potential for long-term preservation. The cyanobacterial microfossil record is robust throughout the Proterozoic (around 2500 to 543 Ma). The record of organic biomarkers—molecules that are highly diagnostic for their parent organisms—is consistent with the microfossil record. For example, only cyanobacteria are known to synthesise 2-methyl bacteriohopanepolyols, which are transformed in sediments to 2-methylhopanes. The latter have now been identified in rocks as old as 2500 to 2700 million years (My) (13, 14). The stromatolitic carbonates that were widespread along continental margins throughout the Proterozoic have long been associated with cyanobacterial communities. Few stromatolites contain identifiable cellular fossils, but the large, Paleoproterozoic (2500 My old and younger) stromatolitic reefs that rival modern reefs in size, architecture, and extent compel the present author to cite their development as firm evidence for oxygenic photosynthesis having become well established by 2500 Ma. Buick also concludes that late Archean stromatolites, observed in 2700-My-old lake deposits, required oxygenic photosynthesis to develop abundantly in environmental settings that lacked evidence of hydrothermal activity.

The molecular oxygen released by photosynthesis leads to additional evidence for the antiquity of oxygenic photosynthesis. Before oxygenic photosynthesis arose, ambient oxygen levels were insignificantly low, because chemical sinks such as reduced geothermal outflows and rock weathering greatly exceeded the main abiotic source of oxygen, the photochemical dissociation of water vapour coupled with loss of hydrogen to space. The discovery of sterane biomarkers in 2700 Ma sediments demonstrates not only the existence of eukaryotic organisms, but also that free oxygen was available for sterol biosynthesis. The extremely low 13C/12C values in 2800-My-old kerogens have been attributed to methanotrophic bacteria, which require both oxygen and methane. The substantial deposition rates of ferric iron in massive banded iron sediment formations before 2500 Ma are clearly consistent with an abundant biological source of free oxygen. Indeed, vast sedimentary deposits of organic carbon, reduced sulphide, ferric iron, and sulphate on continental platforms and along coastal margins are among the most prominent and enduring legacies of billions of years of oxygenic photosynthetic activity.

The geologic record thus offers strong evidence for the evolution of oxygenic photosynthesis before 2800 Ma. There are, however, hints of even earlier origins. The microfossil record of cyanobacteria may extend to 3300 to 3500 Ma, although the evidence for these early Archean occurrences is controversial. Stromatolites occurring in 3460-My-old carbonates and silicified carbonates of the Warrawoona Group, Western Australia, were recently described by Hofman *et al.* These stromatolites developed in a partially restricted, low-energy shallow hypersaline basin. Hofman *et al.* conclude that microorganisms were involved in the accretion of these stromatolites. They also surmise that microbial phototaxis (light-stimulated microbial motility) may have played a role in shaping them, but conclude that the evidence for the presence of photosynthetic biota is not yet definitive.

The carbon isotopic record of early Archean carbonates and reduced carbon is consistent with, but not yet compelling for, oxygenic photosynthesis. The isotopic patterns are consistent with isotopic discrimination by some chemoautotrophic bacteria and anoxygenic photoautotrophic bacteria, in addition to oxygenic photoautotrophs. But the absence of conclusive evidence should not be interpreted as conclusive evidence of absence. Although the early Archean fossil record contains at best a handful of demonstrated microfossils, none of which are as yet physiologically definitive, this low apparent diversity should not be interpreted too literally. It is known that as metamorphic alteration of fossil-containing rocks intensifies, the apparent diversity of remaining microfossil assemblages decreases. Conclusive evidence for oxygenic photosynthesis probably also diminishes. Current evidence for early Archean oxygenic photogenesis is perhaps not yet compelling, but it is consistent with it.

Earth's biogeologic clock: The great antiquity of our biosphere contrasts sharply with the relative youth of plants and animals. The dual geological and molecular biological records of microorganisms indicate that our early biosphere was remarkably complex. All of the major photosynthetic groups of bacteria arose prior to 2800 Ma, perhaps much earlier.

The report by Xiong *et al.* adds an important constraint to the perspective outlined above. The authors demonstrate conclusively for the first time that the major lineages of pigments involved in anoxygenic photosynthesis arose before the development of oxygenic photosynthesis. This indicates that the six major bacterial lineages had largely developed by the mid-Archean, around 3000 to 2800 Ma, and perhaps much earlier. The study also shows that the early biosphere passed through a stage during which even its photosynthetic populations depended exclusively on abiotic sources of reducing power. Can we recognise such a stage in the geologic record? The fossil record of anoxygenic phototrophic bacteria is poorly known, although ancient populations have been identified in much younger rocks on the basis of organic biomarkers. The presence of Chlorobiaceae in Paleozoic sediments has been inferred based on the identification of a 13C-enriched aromatic polyisoprenoid. Geoporphyrins from purple sulphur bacteria have been identified in ancient shales. The survival of traces of Archean oil offers the possibility to extend the biomarker record of anoxygenic phototrophs considerably.

As the great antiquity of photosynthesis becomes more and more apparent, it also becomes easier to envision an ancient, global biosphere sustained principally by anoxygenic photosynthesis. The global geothermal heat flow was substantially higher during Earth's first billion years, and the vigorous geothermal outgassing probably dispersed reduced chemical species throughout sunlight aquatic environments. Perhaps the substantial decline in thermal activity between 4000 and 3000 Ma created opportunities for oxygenic photosynthesis to develop. Both the geologic and living biological records of our early biosphere promise further key insights into the origin and early evolution of photosynthesis.

Molecular Evidence for Early Evolution of Photosynthesis

The origin and evolution of photosynthesis have long remained enigmatic due to a lack of sequence information of photosynthesis genes across the entire photosynthetic domain. To probe early evolutionary history of photosynthesis, we obtained new sequence information of a number of photosynthesis genes from the green sulphur bacterium *Chlorobium tepidum* and the green non-sulphur bacterium *Chloroflexus aurantiacus*. A total of 31 open reading frames that encode enzymes involved in bacteriochlorophyll/porphyrin biosynthesis, carotenoid biosynthesis, and photosynthetic electron transfer were identified in about 100 kilobase pairs of genomic sequence. Phylogenetic analyses of multiple magnesium-tetrapyrrole biosynthesis genes using a combination of distance, maximum parsimony, and maximum likelihood methods indicate that heliobacteria are closest to the last common ancestor of all oxygenic photosynthetic lineages and that green sulphur bacteria and green non-sulphur bacteria are each other's closest relatives. Parsimony and distance analyses further identify purple bacteria as the earliest emerging photosynthetic lineage. These results challenge the conclusions based on 16*S* ribosomal RNA and Hsp60/Hsp70 analyses that green non-sulphur bacteria or heliobacteria are the earliest phototrophs. The overall consensus of our phylogenetic analysis, that bacteriochlorophyll biosynthesis evolved before chlorophyll biosynthesis, also argues against the long-held Granick hypothesis.

Clues to Evolution of Photosynthesis

When early microbes evolved, some species developed ways to convert sunlight into cellular energy and to use that energy to capture carbon from the atmosphere. The origin of this process, known as photosynthesis, was crucial to the later evolution of plants. The analysis of the complete genome sequence of an unusual photosynthetic microbe provides important insights into studies of how that light harvesting mechanism evolved and how it works today.

The bacterium, *Chlorobium tepidum*, was originally isolated from a hot spring in New Zealand. It is a member of the green-sulphur bacterial group, so known because of the microbes colour and their dependence on sulphur compounds to carry out photosynthesis. Biologists say green-sulphur bacteria are important because they perform photosynthesis in a different way from that of other bacteria and that of plants.

For example, instead of the chloroplasts found in plants, green-sulphur bacteria have organelles called chlorosomes that help generate energy through an electron-transport chain in the microbe's cytoplasmic membrane. Inside the chlorosomes, the chlorophyll and carotenoid molecules that capture light differ from the molecules that other species use to perform photosynthesis. Also, green-sulphur bacteria carry out photosynthesis in the absence of oxygen and do not produce oxygen as a by-product as plants do.

> "Because of their unusual mechanisms of harvesting and using the energy of light, the green-sulphur bacteria are important to understanding the evolution and the mechanisms of both photosynthesis and cellular energy metabolism," said Jonathan A. Eisen, an evolutionary biologist at The Institute for Genomic Research (TIGR) in Rockville, Maryland. "The ability to carry out photosynthesis in the absence of oxygen is particularly important to evolutionary studies since it is believed that the early atmosphere of Earth had little oxygen. That is why some scientists have suggested the green-sulphur bacteria were the first photosynthetic organisms."

The sequenced genome of *C. tepidum*, published in the Proceedings of the National Academy of Sciences, represents the first time that a microbe in the green-sulphur group has been fully sequenced. It is also the first time that a bacteria that is both photosynthetic and anaerobic has been sequenced.

Green-sulphur bacteria such as *C. tepidum* are widely distributed in aquatic environments where light reaches anoxic (low-oxygen) layers of water containing reduced sulphur compounds. When TIGR researchers analysed the microbe's single circular chromosome, they identified numerous genes that may play novel roles in photosynthesis or other processes that make use of the energy of light.

Using a method of comparing complete genomes known as phylogenomic analysis, the scientists found strong similarities between the metabolic processes of the green-sulphur bacteria and processes in many species of Archaea, the organisms that represent the third domain of life. This analysis also revealed the likely duplication of genes that are involved in the pathways for photosynthesis and in the metabolism of sulphur and nitrogen. "These duplication events may help explain why this microbe is able to use lower levels of light to carry out photosynthesis than other species," said Eisen, the first author of the PNAS paper.

Another reason why biologists study green-sulphur bacteria is that their mechanism of capturing carbon dioxide differs from that of plants and other bacteria. The green-sulphur bacteria use an unusual chemical cycle called the reductive tricarboxylic acid (TCA) cycle that differs from the Calvin Cycle that is used by higher plants. The TCA cycle uses electrons derived from hydrogen or reduced sulphur compounds to fix carbon dioxide; in contrast, the Calvin Cycle requires oxygen. In fact, the reductive TCA cycle was first discovered in *C. tepidum.*

The authors say that "further genome analysis and experimental work should help provide insights into the evolution of photosynthesis and other pathways of energy metabolism." Scientists have developed methods to genetically manipulate *C. tepidum,* allowing experimental testing of hypotheses generated from the analysis of the microbe's genome. Also, the completion of a genome sequence makes experimental studies easier

Developments

All life on earth is driven by the power of green, as in green chlorophyll. The process at the heart of nature's most efficient engine is photosynthesis. Most of the difficult work occurs in trillionths of a second. ASU researchers are teasing out the details of life's most important chemical reaction.

It takes eight minutes for a photon of light to travel the 93 million miles from the sun to the earth's surface. A green plant needs only a few seconds to capture the energy in that light, process it, and store it in the form of a chemical bond. The amazing process for converting light energy to stored energy is called photosynthesis.

Photosynthesis as a process includes some of the fastest known chemical reactions. The most important events in a photosynthetic reaction occur in trillionths of a second. Measuring such short-lived events, and understanding the chains that link them together, demand some of the most precise experiments and exact measurements technology currently allows.

The stakes are high. Ultimately, almost all life on our planet is fuelled by the power of green—green chlorophyll, that is.

That is one reason why in 1988, the National Science Foundation and the US departments of Agriculture and Energy funded the creation of Arizona State University's centre for the study of early events in photosynthesis. The centre includes more than 20 scientists from the departments of chemistry and biochemistry, botany, and teams of graduate and undergraduate students. All contribute brain power towards unravelling the exact chain of events that occurs in those first trillionths of a second.

ASU researchers use magnets and microwaves, lasers and

spectroscopes, petridishes, ultracentrifuges, and a bevy of sophisticated instruments in their work. To get information, they peer into the infinitesimal nooks and crannies of biological molecules. They watch, they record, they measure, they experiment, they learn.

Learning exactly how some thing or some process works often takes a variety of approaches. One group might devise methods to take that something apart, breaking it down into basic parts. Another group may work to put the parts back together. Still others might study how to modify the parts and then create something new. Studying history often helps. Knowing how something was designed, and the circumstances under which it was created, can help the understanding of its function and structure.

Robert Blankenship is an ASU chemistry professor. He studies the origin and early evolution of photosynthesis. Scientists recently discovered non-photosynthetic bacteria living deep in the earth and under the sea. But there is no direct fossil evidence that these tiny organisms evolved first. "Some people think that photosynthesis existed very near the origin of life on this planet. It's clear from fossil records that some organisms living 3.5 billion years ago were photosynthetic. They are the oldest evidence for any life on earth," Blankenship says.

Scientists have known for some time that photosynthesis is the means by which plants and certain bacteria use sunlight to produce complex organic molecules. The heart of the process is the capture of light energy by small pigment molecules such as chlorophylls, carotenoids, and phycobilins. These molecules are connected along a scaffolding of proteins which are in turn embedded inside a specialised structure. This structure is called the photosynthetic membrane.

Once absorbed by the pigments, captured light energy is then transferred along a chain of molecules to a photosynthetic reaction centre. It is in the reaction centre that the actual chemistry of photosynthesis occurs.

In the reaction centre, the energy from light is used to

separate positive and negative electrical charges across the photosynthetic membrane. The resulting charge-separated state is a form of stored chemical energy. The plant or bacterium uses this energy to power other cellular processes.

The earliest photosynthetic organisms were prokaryotes, living organisms without a central kernel, or nucleus. They were very similar to today's photosynthetic cyanobacteria, also known as blue-green algae.

However, the majority of plants we see around us are eukaryotes. Eukaryotes are trees and corn and wheat. They are seaweed and lily pads, grass and dandelions. Eukaryotic plants are made of billions of cells all working towards shared goals. Each cell includes a nucleus and a slew of other internal machinery called organelles.

Blankenship describes the distinction. "A cyanobacterium has a membrane system inside the cell called a thylakoid membrane. Bacteria do their photosynthesis within the actual cell membrane. In plants, the thylakoid membrane is contained within a specialised organelle called a chloroplast."

Cyanobacteria are very simple organisms. Consider the length of their DNA, which is only about 3.5 million units long. Fully uncoiled, that is about one millimetre in length. In comparison, consider that human DNA contained within a single cell is several metres long.

The long-term goal for Blankenship and other scientists is to better understand how the more complicated photosynthetic machinery of plants evolved from the simpler systems found in photosynthetic bacteria.

Eukaryotic plants probably emerged only a billion years ago. But their emergence required a rather surprising event. In essence, at some point, one cell swallowed the other. But in this case, it was not supper time. The cells ended up working together to form a better, stronger organism. This type of working relationship between two organisms is called symbiosis:

"It's now generally accepted that there was an endosymbiosis, an event in which the cyanobacterium entered into a symbiotic relationship with another cell," Blankenship explains.

"Ultimately, this cyanobacterium became a kind of slave. That is what became of what we now call the chloroplast. If you look at a chloroplast, in effect, you are looking at the remnants of what was once a free-living bacterium."

Blankenship follows the clues suggested by these similarities. He is developing a more precise account of exactly how the various classes of photosynthetic organisms came to be over the last 3.5 billion years.

Higher plants have two types of reaction centres. These centres, known as photosystem I and photosystem II, are connected by the chain of electron carriers. In contrast, bacteria have only one type of reaction centre. But some have reaction centres similar to photosystem I, while those found in others are more like photosystem II.

Photosystem II splits water into oxygen and hydrogen atoms. Photosystem I creates the materials used in later photosynthetic reactions such as food production for the plant.

Both photosystems are found in the chloroplasts of higher plants. They also resemble closely the two photosystems found in the membrane of cyanobacteria. However, all photosynthetic protein complexes have underlying functional and chemical similarities:

"What we find is that there really are just two classes," Blankenship says. "One of them we call the pheophytin quinone. The other is the iron-sulphur type of reaction centre."

Primitive purple photosynthetic bacteria have the pheophytin quinone reaction centre, which chemically resembles the photosystem II of higher plants. Green sulphur and heliobacteria have only the iron-sulphur photosystem, which resembles photosystem I. However, none of these more primitive photosynthetic forms can make oxygen:

> "The really big advance in the history of the earth is the evolution of plants' ability to produce oxygen," Blankenship says. "That is what made it possible for organisms like our ourselves to do respiration. However, the evolutionary source of this activity is still a mystery."

The cellular machinery that constructs proteins from the instructions given in the cell's DNA has a great deal of biochemical similarity—not just across species, but across kingdoms. For example, genetic engineers know that it is possible to insert the human gene for insulin into *Escherichia coli* bacteria. The bacteria then become living factories that can produce large quantities of insulin. Such work is the essence of biotechnology.

But such abilities also give the evolutionary detective new tools with which to dig out bits of information. For example, scientists can now analyse sequences of life's master molecules, DNA and RNA, as well as the proteins that they issue orders and blueprints to build. More importantly, it is possible to assess how long ago, it was that these proteins arose from the same "parent" DNA or RNA.

Blankenship compares the reaction centres in various bacteria and plants alive today. The goal is to determine statistically how closely these reaction centres are related.

The ASU chemist's careful analysis, has challenged some commonly accepted ideas about how these systems evolved. He and ASU botanist Wim Vermaas did experiments to analyse the sequences of the reaction centres in recently discovered primitive photosynthetic organisms called heliobacteria. They were the first to demonstrate that a very primitive complex still exists in nature.

Today, scientists understand fairly well which molecules make up cells, and to a greater extent, how those molecules work. The basic molecular building blocks of cells come in four types: proteins, nucleic acids, lipids, and polysaccharides. Proteins are built from chains of 20 different amino acids, which are assembled in different sequences. Proteins fold into complex three dimensional shapes that provide cellular structure and framework. They also form the machinery that does most of the work.

Nucleic acids come in two forms, DNA and RNA. They store, transfer, and read information. The DNA molecule is made up of two sugar-phosphate strips holding chains of four chemical bases: adenine (A), thymine (T), cytosine (C), and guanine (G). The bases bond (A only with T; C only with G) to join the strips in the famous double-helix structure.

Humans have about 3 billion of these bonded base pairs. Every single human cell contains a yard of DNA, tightly coiled.

Lipids are fatty acids, short strings of atoms that combine to form watertight layers. Lipids protect the cell's insides from the trials and tribulations of the outside environment. Polysaccharides form the cell's energy storehouses.

Every cell is 70 per cent water. Most biological molecules are either hydrophilic (water loving) or hydrophobic (water fearing). Lipids have a water loving end. They orient themselves towards the inside or outside of the cell, and pile back-to-back, to form a watertight layer.

One gene codes for one protein, while every three bases of the gene code for one of the amino acids in the protein. To modify a protein, one has to displace or add an amino acid to the string that makes up that protein. Sometimes, nature does this accidentally by making copying errors in DNA. A string of DNA might look like this: G G C T A A T G C A. But one base might be changed, added, or deleted. The cell's copying mechanisms usually are quite accurate. Mistakes occur only at the rate of 1 in 100,000 to in one million bases.

Genetic engineers work to modify DNA sequences. Their intent is to co-opt the cell into building modified proteins. The challenge is to find the means of inserting modifications that are both accurate and that take hold inside the cell.

Wim Vermaas likes to juggle amino acid residues, the 20 medium-sized molecules which, when connected in millions of configurations, make proteins. He uses a complicated apparatus called an oligonucleotide synthesiser to assist his juggling:

> "This machine makes pieces of DNA," says Vermaas, a professor of botany. "We can design oligonucleotides that are degenerate in regions. That means we can make a zillion different molecules. The catch is that a quarter of them have an A in one position, while another quarter have a C in the same position, another have a G, and the final quarter have a T. We can do that at multiple positions."

The manipulation allows the scientists to have a large number of different DNA sequences represented. Each bit of DNA has the same sequences at places where Vermaas wants them to be the same, but all have different sequences at other places.

Vermaas needs the power of such a technique. He investigates how photosystem II moves electrons across the cell membrane. His approach is a little like Thomas Edison building the light bulb—he tries everything to see what works. His organism of choice is a cyanobacterium.

By creating large numbers of bacterial mutants, Vermaas learns exactly which amino acid residues are important for electron transport, and which residues can be replaced without any significant change in function. Part of this research is guided by intuition, part of it is blind.

Vermaas and his group have made predictions about how changes in the structure of the protein are likely to affect the performance of electron transfer. Their understanding of what is important to molecular function is steadily improving.

Typically, Vermaas modifies DNA sequences of no more than 15 bases at a time. The DNA chains he works with may stretch past 3,000 base pairs. He is able to attach short modified sequences to longer chains of DNA. Vermaas and other scientists routinely use polymerase chain reaction (PCR), a technique for quickly reproducing strands of DNA. He then plugs the modified chains directly into the genome of cyanobacteria.

"PCR lets me amplify DNA very specifically from specific genes," he says. These new strings are amplified in *E. coli* bacteria, and introduced into cyanobacteria. The cyanobacteria have been engineered to lack the functional part of the protein Vermaas is trying to replace.

Cyanobacteria reproduce at stunning rates. As a result, within days, Vermaas has many new varieties of photosystems at his disposal. He tests them using other sophisticated techniques such as electron paramagnetic resonance imaging and laser spectroscopy.

Using laser spectroscopy, scientists fire extremely short pulses of light at the organisms. They then measure how much light is absorbed, how much is reflected, how much fluoresces back out, and how much the spectrum is modified. Vermaas uses the technique to determine the function and efficiency of his newly-created mutants. Machines are used to determine the sequence of amino acids that make up a protein. The process is routine. However, determining the geometry of how an amino acid chain folds is one of the most difficult tasks facing scientists today.

The technique of choice is X-ray diffraction. Scientists try to rebuild a protein's structure based on the pattern of interference generated by sending X-rays through a sample.

James Allen uses X-ray diffraction to study the photosynthetic reaction centre of *Rhodobacter sphaeroidie*, a purple bacterium. X-ray diffraction analysis is very difficult.

> "We have to do what's called a Fourier transform," the ASU chemistry professor explains. "The result is an electron density map. This map tells us the location of all the electrons in the structure. Then we must interpret that information in terms of atoms."

Density map information does not tell the scientist what type of atom is present, only that an atom is there. For Allen to know what is an oxygen and what is a carbon requires knowing the amino acid sequences. In this case, the sequencing was provided by his wife, ASU chemist Joann Williams.

> "It's easy to grow and isolate the protein from this particular bacterium," Williams says. "We did this work in the early 1980s (at University of California, San Diego)."

Teamwork was the key to their lab's success.

> "All of these techniques were just being developed at the same time. That's what made it so powerful," Allen says. "We got the sequence, and genetic work was being done at the same time as the three-dimensional work."

Their achievement is close to unique. "There's only structural information for two reaction centres. This is one of them," Allen adds. The structural knowledge is broadly applicable. Allen says that the reaction centre from purple bacteria is ancestral to photosystem II.

Deciphering the structure of membrane proteins is a dirty problem. Neal Woodbury takes it as a personal challenge.

"In order to pull proteins out of the membrane, we must replace that membrane with something else that acts like a membrane," the chemistry professor explains.

The reaction centre protein structure is enormously complicated. "There are a thousand amino acids," Allen says. "That's more than 10,000 atoms." Woodbury shares laboratory resources with Allen and Williams. The scientists use a technique called site-directed mutagenesis to modify the reaction centres of *R. sphaeroidie* and related bacterial species.

The idea is to modify the structure of the reaction centre protein complex, then see how the modifications affect its function. Allen and Woodbury have somewhat different goals in this pursuit, however.

Allen's group is replacing one amino acid at a time. The change is very specific. They believe they are on their way to understanding how the whole class of photosystem II-type reaction centres functions in general.

Because Allen and his group have established the three-dimensional structure of this reaction centre, much can be built on that knowledge.

> "In our case, we know exactly where all the residues are," Woodbury explains. "We can actually determine what the changes are by using X-ray diffraction. We now can make more sophisticated models."

Woodbury uses site-directed mutagenesis to reach a different goal. "We're interested in why the two sides of the reaction centre are different," he says. "We've been taking pieces from one and sticking it on the other."

The reaction centres changed at several points in evolutionary history. Blankenship's comparative studies show that much. Woodbury wants to understand what functional advantage comes from the two identical parts becoming different. In particular, the structure suggests that there are two pathways for electrons to travel through the reaction centre, yet they take only one path.

Micromachinery

The photosynthetic reaction centres use light energy to drive electron transfer reactions. The goal of this process is to convert the energy from the light into a form that can be used by the living organism. But initially, much of the light energy is used to make the innermost vesicles of the chloroplast much more acidic than their surroundings. Although this does serve as a way to store the energy, it has some severe limitations. First, the protons that cause the acidity rapidly leak out of the vesicles. Energy is lost quickly, much like a car battery that won't hold a charge.

The second problem is that few biological processes in a living organism can tap into this energy source. For example, try to run all your electrical devices designed for 110 volts off of a 12-volt car battery. Without adapters, it cannot be done.

To get around these problems, photosynthetic systems contain a large protein complex called F1F0-ATPsynthase. The protein allows protons to flow through the F0 portion and use the energy to convert ADP (adenosine diphosphate) and phosphate into ATP (adenosine triphosphate). Think of ATP as high test gasoline for cells. ATP is very stable, making it ideal for energy storage.

ASU chemist Wayne Frasch studies the ATPsynthase enzyme. He is excited. Scientists recently completed the X-ray crystal structure of the enzyme's F1 portion. This portion is responsible for ATP synthesis. Frasch says the portion looks like an orange with six protein subunit slices that surround another protein subunit. Each of the slices contains a binding site for ATP, but only three of them are responsible for the conversion of ADP and phosphate into ATP. The function of the other three slices remains unknown. "Each of the slices has what appears to be a lever at the bottom. The structure is provocative in that it appears that the central subunit might spin around the inside, perhaps in response to the movement of protons," Frasch explains.

As it spins, this central unit might raise and lower the levers in a manner that might cause ATP to be formed. "If correct, the enzyme would be a mechanically-driven, molecular pump!" he adds.

X-ray crystal structures provide only a snapshot view of something that must move in many complex ways to function. Frasch is able to follow many of these important structural changes through a combination of EPR spectroscopy and genetic engineering using the gene gun on *Chlamydomonas*.

The enzyme uses a metal atom, usually magnesium, to bind the phosphate and the phosphate groups of ADP in order to make the bond between them. By substituting the magnesium for vanadyl, which gives rise to an EPR signal, the changes that occur to the metal as ATP is made can be followed directly. "Although we can't see all the changes that take place, it's like shining a flashlight on the very spot you want to see," Frasch says.

When amino acids crucial to this process are changed by genetic engineering, the changes are easily seen in the EPR spectra from the vanadyl bound at the site. In this way, many of the subtle but important mechanistic features of the enzyme are coming to "light" (fitting for a photosynthetic process).

Getting Past Cellulose

Most of the scientists at ASU's Photosynthesis Centre work on bacterial photosystems for very practical reasons. Bacteria are small, so small that millions can live inside an ordinary beaker. They reproduce so quickly that one organism can become a billion in less than a week. Bacteria also contain a limited amount of DNA. New DNA can be inserted relatively easily by using sexual conjugation with ordinary *E. coli* bacteria.

But eukaryotic plants, the photosynthetic organisms that affect us directly, are more complex. Andrew Webber studies one of these eukaryotes. The plant is a little simpler than some, but shares most of the important biochemical features of plants that cover hills and valleys.

Chlamydomonas reinhardtii is a one-celled algae. Algae are similar to higher plants. Their cell walls contain a separate chloroplast. The chloroplast contains only 196,000 base pairs of DNA. Many of the proteins necessary for photosynthesis must be assembled outside the chloroplast and imported.

Webber studies photosystem I in *C. reinhardtii*. Doing the sort of genetic engineering so popular among his colleagues involves special challenges. Actually, it involves a gun. A biolistic gene gun, to be exact.

C. reinhardtii is a plant, not a bacterium. It has tough cellulose cell walls, just like the plants growing outside your window. Typical methods for inserting genetic material, such as bacterial conjugation, just won't work.

For one thing, bacteria show no great efforts to mate with an organism from another kingdom. For another, the cell wall would probably prevent any such attempts.

"For years and years, people have been trying to refine techniques for transforming organelles," Webber explains. "The gene gun is a method that works." Scientists mix purified engineered DNA with very small balls made of tungsten or gold. The balls are less than a micron in diameter. Consider that the smallest grain of sand on a beach is about 90 microns in diameter. The gene gun is fired using compressed helium.

At sufficient helium pressure, a plastic membrane ruptures and fires the particle mixture down a barrel at the *C. reinhardtii* target. Success at creating mutants results through selection and reproduction.

> "It is a numbers game," Webber explains. "That is why we work on *C. reinhardtii*. We can grow a large number of cells, and it has the advantage of having only one chloroplast per cell."

The chloroplast itself is huge; taking up 70 per cent of the cell's volume.

> "When we shoot *C. reinhardtii*, there's a good chance that the particle will go into the chloroplast as opposed to anywhere else."

Inside the cell there are perhaps 80 copies of chloroplast DNA, and the new DNA reaches perhaps only one of those. But the new DNA is special. It carries a gene for antibiotic resistance. Webber simply kills off the algae that don't contain the new gene.

Over time, numerous bouts of antibiotic treatment leave only algae that possess many copies of Webber's gene for antibiotic resistance, and thus carry as well his genetic changes.

Webber wants to better understand the mechanisms of the photosystem I reaction centre in higher plants, particularly the function of co-factors like the iron-sulphur complex and reaction centre chlorophylls, which are essential for electron transfer.

Historical Timeline

384-322 (BC): *Aristotle* compared the soil (earth) to the stomach and concluded that the earth is like the stomach of plants as they gain their nutrients directly from earth and water without having a "proper" digestive system.

1648: *Johan Baptist van Helmont* (from Brussels) considered water to be the source of life and the basic nutrient for plants. Therefore he devised an experiment by which he showed that small potted willows can thrive on soil and water alone while they gain their substance (weight) solely from the "water" as the weight of the soil in the pots did not decrease significantly. He published his conclusion in his book "Ortus medicinae" wherein he also used the term gas for the first time.

1675: *Marcellus Malphigi* (from Bologna) was the first to study the anatomy of plants (and insects) concisely by making use of the microscope and he claimed that plants take up nutrients which are dissolved in water via their roots.

1679: In a letter to a friend which was later published in 1717 *Edme Mariotte* (from Dijon) summarised the "current" knowledge about the composition and nutrition of plants and thus provided the first theory of metabolism by stating that all plants are made up of certain basic substances (like sulphur, salt, oil, ammonia and nitre) which must be contained in the soil, where from they are consequently taken up by the plants and transformed into their metabolites. As plentiful different plants can grow from the same soil and water he concluded that each plant must provide its own specific metabolism in order to create its substances and form.

1684: In this year the *Irishman Robert Boyle* who was a prominent proponent of the Corpuscular Philosophy published his "Memoirs for the Natural History of *Human Blood"* in which he documented that the changing colour of blood during its passage through the capillaries of the lung (from dark to light red) was due to a certain ingredient of air which was consumed by fire or breathing.

1703: The German *G.E. Stahl* (later the physician of Friedrich Wilhelm I. in Berlin) put forward his *Phlogiston Theory* according to which an element called phlogiston was given off by combustible materials when they burned. Air in which things had been burned became less able to support combustion because, it was thought, it was saturated with phlogiston which, according to the current belief, could then also mix with other substances. It was assumed that phlogiston ought to have a negative weight because substances which had been burned were lighter than before.

1727: *Stephan Hales* (an English clergyman) described the leaves as organs of transpiration and he postulated that plants exchange gases with their surrounding air. Furthermore, he was the first to point out a possible role of light in plant nutrition by quoting Newton: *"Könnte nicht eine beiderseitige ineinander wirckende Verwandlung (transformation reciproque) zwischen dicken Körpern und Lichte vorgehen ? und die Körper einen grossen Theil ihrer Activität von denen Lichtparticuln haben, die in ihren Zusammensatz mit kommen. Daß Körper in Licht und Licht in Körper verwandelt werde, ist doch der Natur Lauf gar gemäß, welche am liebsten mit Verwandlungen (transformations) zu schaffen hat."*

1772: *Joseph Priestley* (from Yorkshire) published his

"Experiments and Observations on different kinds of air" and was the first to prove the different qualities of the gases released by plants and the one´s exhaled by animals (mice). He discovered that, although a candle burned out in a closed container, when he added a living sprig of mint to the container, the candle would continue to burn. At the time, Priestley did not know of O_2, but he correctly concluded that the mint sprig "restored" the air that the burning candle (or mice which he used in a similar set of experiments) had depleted.

As he still believed in the *Phlogiston Theory* he called the gas which was given off by plants dephlogisticated air because a candle would burn brightly in it.

1779: *Jan Ingenhousz* (from Breda, Netherlands): "Experiments upon vegetables, discovering their great power of purifying the common air in the sunshine and of injuring it in the shade and at night, to which is joined a new method of examining the accurate degree of the atmosphere" systematically investigated the release of dephlogisticated air from green parts of plants during day time especially from the lower side of leaves, while he also discovered that this depends on the effects of light as plants give off noxious air during night time.

1783: Priestley divulged his discovery personally to the eminent French chemist *Antoin-Laurent Lavoiser* who immediately understood the theoretical implications of it, as Priestly did not. Lavoiser had already announced, in 1772, that he was destined to bring about a revolution in physics and chemistry. Unlike the older scientists he realised that atmospheric air was not an element but a compound of gases, and he identified Priestley´s discovery as the active component of air for which he had been searching. He called it oxygen (Greek: acid former), in the belief that all acids contained it.

1796: After *Ingenhousz* learned about the findings of Lavoiser he elaborated on his on investigations and stated that green plants absorb carbon dioxide and release oxygen during day time.

1793-98: *Hedewig, Schrank and Humboldt* discuss the function of stomata.

1804: *N. Theodore de Saussure* (Switzerland): "Recherche chimiques sur la *végétation*". He verified Ingenhousz´s hypothesis that plants assimilate carbon dioxide from the air while nitrogen and other nutrients are derived from the soil. Furthermore he was the first to distinguish systematically between the principles of assimilation and dissimilation.

1817: *Pelletier and Caventou* isolated the green substance in leaves and named it chlorophyll.

1840: *Liebig´s* findings help to establish the use of fertilizers in agriculture.

1845: *Robert Mayer* plants transform energy of sunlight into chemical energy. Energy can be transformed but not created or destroyed (first law of thermodynamics or energy conservation).

1845: *Mohl* Discovery of starch associated with chlorophyll.

1851: *L. Garreau* again delineated the differences between dissimilation and assimilation and emphasises that dissimilation continuously takes place in all parts of a plant.

1864: *Bossingault* made the first accurate measurements of the gas exchange and found that the volume of O_2 evolved and CO_2 used up is almost unity.

1862-64: *Julius Sachs* (from Breslau) investigated the synthesis of starch under the influence of light and in relation to chlorophyll. He worked out the overall equation of photosynthesis:

$$6\,CO_2 + 12\,H_2O + \text{solar energy} \rightarrow C_6H_{12}O_6 + 6\,O_2 + 6H_2O$$

1869-71: *K.A. Timirjazev* (from St. Petersburg) developed a method to investigate the spectrum of light and against the background of the law of energy conservation he stressed the pivotal role of green plants for the transformation of light energy into chemical energy.

1872: *W. Pfluger* defined respiration as a process located on the cellular level.

1883: The replication of chloroplasts was observed by *F. Schmitz and A.F.W. Schimper.*

1883: An elegant experiment was first performed by the German botanist *Thomas Engelmann.* He illuminated a filamentous alga with light that had been passed through a prism, thus exposing different segments of the alga to different wavelengths of light. Engelmann used aerobic bacteria, which seek oxygen, to determine which segments of the plant were releasing the most O_2. Bacteria congregated in greatest density around the parts of the alga illuminated with red and blue light.

He also demonstrated the correspondence between the action spectrum of *photosynthesis* and the absorption spectrum of chlorophyll.

1887: *Sachs* reviewed the current knowledge and concluded: "the most definite proof that ... the chlorophyll body chloroplast itself is the organ which decomposes carbon dioxide and consequently assimilates this organic substance, is afforded by the fact ... that the first recognisable products of assimilation starch appear not in any haphazard place in the green cell, but in the chlorophyll body itself." Until the findings of Hill and van Niel scientists now believed that chloroplasts are the site of complete photosynthesis because it was still thought that oxygen evolution and CO_2 assimilation were inseparable events.

1888: *Pfeffer* "Handbuch zum Stoffwechsel und Kraftwechsel in der Pflanze"

1900: *R.Hober* establishes the measurement of pH-values.

1901: *K.A. Timirjazev* studied drought tolerance and was the first to reveal the antagonism between transpiration and assimilation.

1902: *Max Rubner* was able to prove that the first law of thermodynamics (energy conservation) is also applicable to organisms ("Gesetze des Energieverbrauchs im Organismus").

1905: The British plant physiologist Blackman interpreted the shape of the light-saturation curves by suggesting that photosynthesis is a two-step mechanism involving a photochemical or light-dependent reaction and a non-photochemical or light-independent reaction.

1895-06: Until the end of the 19th century scientists still assumed that the synthesis and degradation of intracellular substances would require the "living" protoplasm of intact cells. However, this changed with the discovery of isolated biological enzymes which were still active (*Bertrand*, 1895 and *Buchner*, 1897). Thus, the enzyme theory of metabolism was established and the field of Plant Biochemistry developed rapidly. Among the founding fathers were *F. Czapek* (Biochemie der Pflanzen, 1905) and *F.F. Blackman* (1906) who wrote about the function of the protoplasm: "It is a complicated congeries of catalytic agents, adapted to the metabolic work that the cell has to do".

1905: *F.F. Blackman* and *G.L.C. Matthaei* investigated the influence of the factors light and temperature on the assimilation of CO_2 and were able to distinguish between a temperature-dependent but light-independent and a light-dependent but temperature-independent reaction. They suggested a two-step mechanism involving a photochemical or light reaction and a non-photochemical with a high temperature coefficient (indicative of an enzymatic reaction).

1905: *K.S. Merezkovskij* first claimed an endosymbiotic origin of chloroplasts.

1905-36: Between 1905 and 1916 *Albert Einstein* published his theory of relativity while the "uncertainty principle" was formulated in 1927 by the German *Werner Heisenberg*, and consequently the mathematical formulation of quantum theory was developed in the preceding years which also enabled physicists to calculate and predict the emission of energy from the sun. Accordingly the sun is considered to be a giant thermonuclear reactor. The energy it emits comes from fusion reactions much like those that occur in a hydrogen bomb. Four hydrogen atoms fuse to form one helium, which has a mass less than the total mass of the hydrogen atoms. The lost mass is converted to energy, and by means of Albert Einstein´s famous formula ($E=mc^2$) it was calculated that each minute about 120 million tons of solar matter are converted to a colossal amount of energy that radiates out into space. A small fraction of that energy reaches earth in the form of electromagnetic energy (waves/particles), taking only a few minutes to make the trip (the speed of light is about 300000 km/sec).

1906: *M.S. Cvet or Tswett* from Italy introduced chromatography by separating the pigments of leaves.

1909: *Michaelis* first invented electrophoresis.

1909: *C. Correns* identified inheritable factors in plastids.

1913: *R. Willstatter* determined the overall structure of chlorophyll.

1915: *Dixon* established his hypothesis on transpiration and the ascent of sap in plants ("Kohasonstheorie").

1923: *Hevesy* was the first to use to apply the tracer technique to plant physiology as he followed the passage of radioactive isotopes through plants.

1926: *E.Munch* studied source-sink relation and proposed his hypothesis ("Druckstromtheorie").

1926: *Warburg* postulated that CO_2 molecules remain adsorbed on the chloroplast surface until successive step-wise reduction by light-activated chlorophyll converts them to glucose and liberates oxygen.

1927: *Osterhout* is the first to measure a membrane potential.

1927: *Warburg and Negelein* developed spectrophotometry.

1929: ATP described by *K. Lohmann.*

1930: Prior to about 1930, many investigators in the field believed that the primary reaction in photosynthesis was splitting of CO_2 by light to carbon and O_2, while the carbon would subsequently be reduced to carbohydrates by water in a different set of reactions.

1930-41: It was found that some bacteria can assimilate CO_2 and synthesise carbohydrates without the use of light energy. Subsequently, the Dutch microbiologist van Niel who worked mostly in California showed that some bacteria can assimilate CO_2 in light without evolving O_2. After manifold experiments with these sulphur bacteria *C.B. van Niel* concluded that the basic principle of photosynthesis was a light-driven exchange of hydrogen from a donator (H_2A) which was to be oxidised to CO_2 (acceptor) which would consequently be reduced. On the basis of this he postulated that in plants the hydrogen had to be derived from the splitting of water. Formerly, many investigators believed that the primary reaction in photosynthesis was splitting of CO_2 by light to Carbon and O_2.

1932: *Emerson* and *Arnold* measure the length of the light-independent reaction with light flashes lasting less than a millisecond. Flash saturation occurred in normal cells when one molecule of O_2 evolved from 2500 chlorophyll molecules. Emerson and Arnold concluded that the maximum yield of photosynthesis was not determined by the number of chlorophyll molecules capturing the light but by the number of enzyme molecules that carry out the light-independent reactions.

1935: *Danielli* and *Davson's* first model of a membrane containing proteins.

1937: *R. Hill* successfully isolated chloroplasts and separated them from the respiratory particles (the first mitochondria were purified in 1951 by A. Millerd) by differential centrifugation. By using the affinity of haemoglobin to oxygen he was able to measure the release of oxygen from isolated chloroplasts by the use of spectroscopy in the presence of suitable electron acceptors (oxidants) which became reduced. This light-driven transfer of electrons from water to non-physiological oxidants is now known as the Hill reaction. He proved that chloroplasts were able to produce oxygen (split water) in the absence of CO_2. However, the O_2-evolving chloroplasts prepared by Hill failed to photoreduce CO_2, the "natural" electron acceptor of photosynthesis. It was then thought that CO_2 assimilation is impossible in a cell-free system.

1937: *H. Krebs* the citric acid cycle.

1938: *R. Hill* observed the splitting of water with isolated chloroplasts.

1939: *H. Fischer* and *W. Wenderoth* determined the chemical structure of the chlorophylls and their conjugated double bonds.

1939: *Kamen* and *S. Ruben* and co-workers established the decomposition of H_2O and the resulting liberation of O_2 by using $^{18}O_2$. Furthermore, they were the first to use the tracer technology for work with $^{14}CO_2$.

1925-53: A number of other "enabling" techniques for research in plant physiology like climatised growth chambers, ultracentrifugation, electron microscopy, x-ray-diffraction, TL- and GL-chromatography, fluorescence spectrophotometry or the use of aphids to study the phloem became available.

1941: *Fritz Lippmann* and *Hermann Kalckar* defined the general metabolic function of ATP.

1943: The first comprehensive hypothesis for a role of ATP in photosynthesis was formulated by *Ruben* in 1943 who postulated that photosynthetic CO_2 assimilation is a dark process, dependent on photochemically generated ATP and reduced pyridine nucleotides. However, at the time there was no experimental evidence for this and Ruben's hypothesis did not fare well in the succeeding years. It was deemed incorrect on both theoretical and experimental grounds.

1944-45: *R.L. Emmerson* also postulated a key role for ATP during photosynthesis based on experiments with *Chlorella*: "The sole function of light energy in photosynthesis is the formation of energy-rich phosphate bonds." This proposal was strongly opposed by *Rabinowitch* (1945) who found the supporting evidence inadequate and considered the idea theoretically unsound: "What good can be served by converting light quanta (even those of red light, which amount to about

43 kcal per Einstein) into phosphate quanta of only 10 kcal per mol ? This appears to start in the wron direction - towards dissipation rather than towards accumulation of energy."

1944: *R.Consden* and co-workers adapted paper chromatography to the analysis of amino acids and it is now also becoming useful for other compounds.

1945-57: In the year 1945 *Melvin Calvin,* Benson and colleagues had started a series of investigations that resulted in the elucidation of the light-independent reactions of photosynthesis. They used the unicellular green alga *Chlorella* in their work because it was easy to let these organisms grow consistently. Their findings later proved to be pertinent to a wide variety of organisms ranging from photosynthetic bacteria to higher plants. The aim of their work was to determine the pathway by which CO_2 becomes fixed into carbohydrate. The experimental strategy was to use radioactive carbon (^{14}C) as a tracer. $^{14}CO_2$ was injected into a suspension of algae that had been carrying out photosynthesis with normal CO_2. The algae were killed after a certain time (5-60 seconds) by dropping the suspension into alcohol. The cells were homogenised and the radioactive compounds in the algae were separated by paper chromatography. The paper chromatogram was then pressed against photographic film, which became black where the paper contained a radioactive spot.

In his Nobel Lecture (1957), *Calvin* noted that their primary data resided "in the number, position, and intensity — that is, radioactivity — of the blackened areas on paper chromatograms. The paper ordinarily does not print out the names of these compounds, unfortunately, and our principal chore for the succeeding ten years was to properly label those blacked areas on the film."

The radiochromatogram of the algal suspension after sixty seconds of illumination was so complex that it was not feasible to detect the earliest intermediate in the fixation of CO_2. However, the pattern after only five seconds of illumination was much simpler. In fact, there was just one prominent radioactive spot, which proved to be 3-phosphoglycerate. The formation of 3-phosphoglycerate as the first detectable radioactive intermediate suggested that a two carbon compound is the acceptor for the CO_2. This proved not to be so. The actual reaction sequence is more complex. The CO_2 molecule condenses with ribulose 1,5-bisphosphate to form a transient six carbon compound, which is rapidly hydrolysed to two molecules of 3-phosphoglycerate.

This reaction is driven by a protein called ribulose 1, 5-bisphosphate carboxylase (commonly called Rubisco). Rubisco is located on the stromal surface of the thylakoid membrane and it is very abundant in chloroplasts, comprising more than 16 per cent of their total protein. In fact, Rubisco is probably the most abundant protein in the biosphere, and it is one of only two proteins in the world which are able to bind CO_2 and thus fix it into organic molecules The first detectable product of this reaction is 3-phosphoglycerate.

1948, current knowledge of the occurrence of quinones in the photosynthetic apparatus began with the demonstration that the vitamin K_1 of leaves is localised in chloroplasts by *Dam*.

1948-52: The first experiments of *Aronoff* and *Calvin* (1948) with the sensitive ^{32}P technique used to test the ability of isolated chloroplasts to form ATP (on illumination) gave negative results and seemed to contradict the hypotheses of *Rubens* and *Emmerson* (s. above). The most plausible model for ATP formation in photosynthesis became one that envisaged a collaboration between chloroplast and mitochondria. Chloroplasts would, in that scheme, reduce NAD^+ photochemically and mitochondria would reoxidise it with oxygen and form ATP via oxidative phosphorylation according to the model of *Vishniac* and *Ochoa* (1952). But this model also posed a serious physiological problem as photosynthesis in saturating light can proceed at a rate almost 30 times greater than the rate of respiration. It was difficult to see, therefore, how the respiratory mechanisms of mitochondria could cope with the ATP requirement in photosynthesis.

1948-56: *Brown* and *Frank* (1948) thought they had proved by experiments with $^{14}CO_2$ that isolated chloroplasts were not able to fix CO_2 and together with the findings of van Niel and Hill that the photoproduction of oxygen by chloroplasts was basically independent of CO_2 assimilation this led scientists conclude that the chloroplast was an: "incomplete machine ... which can evolve O_2 from H_2O in light provided an external substance B is present ... and that this requirement for B can be attributed to a deficiency in the enzymatic mechanism operative in various steps by which CO_2 is converted into the normal assimilation product." (*Van Niel*, 1956). A few years earlier, when the oxygen evolution by chloroplasts was linked (via $NADP^+$) to CO_2 assimilation (s.a.) by a malic enzyme of cytoplasmic origin in 1951 *Arnon* similarly concluded that "isolated chloroplasts lack either the appropriate hydrogen carriers or enzymes, or both, required for the reduction of carbon dioxide by the hydrogen derived from the photolysis of water, and that in the intact cell these factors are found chiefly or wholly outside the chloroplasts." In sum, it contrast to the prediction of Sachs in 1887 (s.a.) it now seemed well-established that the chloroplast was indeed a "system much simpler than that required for photosynthesis" and was the site of only "the light absorbing and water splitting reactions of the overall photosynthetic process " as *Lumry* put it in 1954. Thus at mid-century, the concept of CO_2 assimilation by chloroplasts was abandoned (for a few months/years).

1949: *Katz* was the first proponent of the idea that photon energy is used in photosynthesis to transfer electrons rather than cumbersome atoms.

1950: The Society for Experimental Biology held symposium on carbon dioxide fixation and photosynthesis at Sheffield, England, and the Proceedings of that Symposium (published in 1951) provide a reliable record of some of the main experimental and conceptual realities that characterised photosynthetic research up to the mid-twentieth century.

1951: *Calvin* finally proved that phosphoglycerate is the first product of carbon dioxide assimilation.

1951: The photosynthetic reduction of NADP to $NADPH_2$ by isolated chloroplasts with simultaneous evolution of O_2 was reported in 1951 by three different laboratories, including *Arnon's*. Thus the first link between oxygen evolution by chloroplasts (via $NADP^+$) to CO_2 assimilation had been found, although the significance of this was not clear to the majority of researchers.

1951: *Hill* confirmed that the photoproduction of oxygen by chloroplasts was basically independent of CO_2 assimilation. In 1951 Hill wrote: "If we break the green cell, it is possible to separate the fluid containing the chloroplast and chloroplast fragments from the tissue. This green juice can no longer assimilate carbon dioxide, but in the case of many plants the insoluble material, for a time at least, is still capable of giving oxygen in light. This evolution of oxygen takes place in the presence of soluble substances contained in the plant juice or better by the addition of reagents which can act as hydrogen acceptors. The cell preparations then though not showing photosynthesis, can still convert light into a form of chemical energy. We may very well call this the chloroplast reaction." This chloroplast reaction became to be known as *Hill Reaction*.

1951: *Burk* and *Warburg* divided the photosynthetic energy conversion process into two parts: (1) a one-quantum light reaction which liberates oxygen and converts a bound species of CO_2 into carbohydrate; and (2) a dark oxidative reaction which provides the rest of the total energy needed for CO_2 assimilation. However, subsequent work with isolated chloroplasts soon yielded evidence that one-quantum light reactions in photosynthesis are concerned not with CO_2 assimilation and terminal events in oxygen evolution but with intermediate light-induced electron transfer steps.

1952-73: Cytochromes were described and characterised as electron carriers by a number of research groups (among them: *Hill, Lundegard, Arnon, Anderson and Bendall*).

1953: In the early 1950s, when it appeared that isolated chloroplasts did not assimilate CO_2, photosynthesis came to be regarded, like fermentation in the days of Pasteur, as a process that could not be separated from the structural and functional complexity of whole cells. Thus, *Rabinowitch* (1953) concluded that "the task of separating it photosynthesis from other life processes in the cell and analysing it into its essential chemical reactions has proved to be more difficult than was anticipated. The photosynthetic process like certain other groups of reactions in living cells seems to be bound to the structure of the cell. It cannot be repeated outside that structure."

1954: *Frenkel* observed the formation of ATP from ADP and Pi on illumination of membrane preparations.

1954: The principle of photophosphorylation (i.e. light-induced synthesis of ATP from ADP and Pi) was proposed and demonstrated by *Arnon*. Isolated chloroplasts were thus able to convert light energy into chemical energy and trap it in the pyrophosphate bonds of ATP. Several unique features distinguished this photosynthetic phosphorylation (photophosphorylation) from substrate-level phosphorylation in fermentation and oxidative phosphorylation in respiration: (1) ATP formation occurred in the chlorophyll-containing lamellae and was independent of other enzyme systems or organelles; (2) no energy-rich substrate, other than absorbed photons, served as a source of energy; (3) no oxygen was produced or consumed; (4) ATP formation was not accompanied by a measurable electron transport involving any external electron donor or acceptor.

The next objective in this avenue of research was to explain the mechanism of photophosphorylation as it seemed unlikely that light was directly involved in the formation of ATP itself because ATP formation was already known to be a reaction which universally occurred in all cells independently of photosynthesis. Light energy, therefore, had to be used in photophosphorylation before ATP synthesis and in a manner unrelated to CO_2 assimilation or oxygen evolution and the most probable mechanism for such a role seemed to be a light-induced electron flow (*Levitt*) which was known to be involved in the ATP formation in non-photosynthetic cells.

1954: *D.I. Arnon* and co-workers provided direct evidence for the long standing hypothesis that CO_2 assimilation takes place in chloroplasts.

1954-56: *Arnon, Trebst, Losada* and *Tsujimoto* succeeded in distinguishing light-dependent from light-independent reactions by separating the chloroplasts into a granal and stromal fraction and demonstrating ATP and $NADPH_2$ formation in the grana in light, and the light-independent CO_2 reduction by enzymes of the stroma.

Later the group discovered cyclic photophosphorylation, which produced only ATP, and proposed its role for providing additional ATP for CO_2 fixation in chloroplasts. For this reason, prior to the discovery of non-cyclic photophosphorylation (1957-61), cyclic photophosphorylation was regarded as the only source of ATP for CO_2 assimilation.

1955: *Arnon* the ability of whole isolated chloroplasts to carry out the synthesis of starch from $^{14}CO_2$ and water without the aid of external enzyme systems and organic substrates provided incontrovertible evidence for concluding that chloroplasts are indeed, as was asserted but not documented in earlier periods the sites of both O_2 evolution and CO_2 assimilation.

1956: The sraw-coloured chloroplast extract (stroma) obtained after osmotic shock treatment of chloroplasts was found to contain all the necessary components for CO_2 fixation (*Whatley*).

1956: The key role of pyridine nucleotides was confirmed by *A. San Pietro* and *H.M. Lang*.

1956: The existence of photochemical reaction centres was proved for bacteria and higher plants (*L.N.M. Duysen* and *B. Kok*).

1957: *R. Emerson* confirmed the existence of two photosystems in the thylakoid membrane which was later verified by *Doring et al.* in 1967.

1957-61: *James* and *Das* (1957) and *Lundegard* (1961) confirmed that chloroplasts cannot respire and lack the terminal enzyme of respiration (cytochrome oxidase). This feature ensured that which was generated had to be produced photochemically and not formed by respiration.

1957-61: *Arnon* and co-workers (1957) isolated a soluble protein, first named the TNP-reducing factor, which when added to low concentrations of broken chloroplasts (grana) catalysed the photoreduction of substrate amounts of $NADP^+$ and not NAD^+. This facto later become recognised as a synonym for ferredoxin, an electron carrier protein that interacts with the enzyme ferredoxin-NADP reductase. However this reduction was first thought to be wholly unrelated or even antagonistic to photophosphorylation. It was therefore a surprise when the group of *Arnon* provided direct experimental evidence for a coupling between photoreduction of $NADP^+$ and the synthesis of ATP in 1958. ATP formation was stoichiometrically coupled with a light-driven transfer of electrons from water to $NADP^+$. It was envisaged that a chlorophyll molecule excited by a captured photon transfers an electron to $NADP^+$. It was postulated that electrons thus removed from chlorophyll are replaced by electrons from water with a resultant evolution of oxygen. In this manner, light would induce an electron flow from OH^- to $NADP^+$ and a coupled phosphorylation. Because of the unidirectional or non-cyclic nature of this electron flow, this process was named non-cyclic photophosphorylation (*Arnon*, 1960 and 1961).

1958: The experiment of *Trebst, Tsujimoto* and *Arnon* (1958, Nature 182) directly substantiated the validity of the concept proposing a light-dependent and a light-independent phase of photosynthesis through a physical separation of the two phases after fractionating isolated chloroplasts. First, they were able to show that chloroplasts produced oxygen, ATP and ATP in a light-dependent manner in the absence of CO_2. Then, after the lamellar (membranous) portion had been discarded and the remaining chlorophyll-free stroma was provided with CO_2, ATP and NADPH the fixation of CO_2 proceeded in a light-independent manner.

1959: *Robertson's* concept of a unit membrane.

1959-61: A hypothesis was put forward by *Arnon* that a chlorophyll molecule, on absorbing a quantum of light, becomes excited and promotes an electron to an outer orbital with a higher energy level. This high-energy electron would then be transferred to an adjacent electron acceptor molecule, a catalyst (A) with a strongly electronegative oxidation-reduction potential. Therefore, the transfer of an electron from excited chlorophyll to this first acceptor would be the proper energy conversion step and thus terminate the photochemical phase of the process. By transforming a flow of photons into a flow of electrons it would constitute a mechanism for generating a strongly electronegative reductant at the expense of the excitation energy of chlorophyll. Several of the exergonic electron transfer steps, particularly those involving cytochromes, were thought to be coupled with phosphorylation. At the end of one cycle, an electron would be returned to the electron-deficient chlorophyll molecule and the quantum absorption process could be repeated.

1960: The groups of *Losada* and *Trebst* identified and characterised the CO_2—fixing enzymes.

1960: *Emerson* and co-workers in Illinois measured the quantum yields of photosynthesis in algae and observed that the average quantum yield obtained by using two superimposed light beams of different wavelengths was higher than the average quantum yield obtained by using the two beams separately. To explain this enhancement of quantum yield, *Emerson* and *Rabinowitch* in 1960 postulated the existence of two light reactions in photosynthesis.

1960: R. *Hill* and F. *Bendall* reviewed the photochemical and thermodynamic studies of the preceding years and put forward the Z-Scheme of electron transport showing the operation of two photosystems, in a series, in photosynthetic electron transport and phosphorylation. These led to the development of the idea of two photosystems operating in a series; PS II in which O_2 evolution occurs, and PS I in which ferredoxin is photoreduced and $NADPH_2$ is formed.

1960-61: Plastocyanin, a copper-containing protein, was discovered by *Katoh* in *Chlorella* and later in leaf tissue and chloroplasts.

1961-62: *Losada* and co-workers experimentally separated non-cyclic non-cyclic photophosphorylation by isolated chloroplasts into two light reactions: one light reaction photo-oxidised water, yielded oxygen, and reduced dichlorophenol indophenol (DCIP).

A second light reaction gave a photoreduction of $NADP^+$ by reduced DCIP and a coupled photophosphorylation. *Duysens* and *Amesz* proposed that photosynthesis in higher plants consisted of two light reactions operating in series: a long-wavelength, photosystem I, which oxidises cytochrome *f* and photoreduces $NADP^+$, and a short-wavelength, photosystem II, which reduces cytochrome *f* and produces oxygen by dehydrogenation of water.

1961-66: *P. Mitchell* (Nature 191, p.144-148) published his work on "Coupling of phosphorylation to electron transfer by a chemiosmotic type of mechanism".

Even after the mechanisms of carbon fixation via ribulose 1,5-bisphosphate was known the fate of protons (H^+) released after the splitting of water and the exact mechanism for the synthesis of ATP via the usage of high energy electrons from the electron transport chains remained an enigma for a long time. For some time scientists assumed that excited electrons might be trapped by an activated protein which could then use their power to drive the synthesis of ATP.

Investigators in many laboratories tried for several decades to isolate these putative energy-rich proteins, but none were found. A radically different mechanism, the chemiosmotic hypothesis, was postulated by Peter Mitchell in 1961. He proposed that electron transport and ATP synthesis are coupled by a proton gradient across the membrane rather than mediated by an activated protein.

According to this model the splitting of water and the electron transport chains induce an increase of the H^+-concentration in the thylakoid space. As a consequence a proton gradient across the thylakoid membrane is generated and finally the movement of protons along this gradient (back to the stroma) can be used to synthesise ATP. Thus, Peter Mitchell speculated that a so called *proton-motive force* drives the activity of the ATP synthase.

A revealing experiment which was performed by *Andre Jagendorf* in the year 1966 provided strong support for Mitchell's hypothesis that ATP synthesis is driven by a proton-motive force. Jagendorf was able to create an artificial pH gradient across thylakoid membranes of isolated chloroplasts. In order to achieve this he first soaked isolated chloroplasts in a pH 4 buffer for several hours. These chloroplasts were then rapidly mixed with a pH 8 buffer containing ADP and P. As a result of this the pH of the stroma suddenly increased to 8, whereas the pH of the thylakoid space remained at 4. A burst of ATP synthesis then accompanied the disappearance of the pH gradient across the thylakoid membrane. Thus, the thylakoid membrane makes ATP as hydrogen protons diffuse from the thylakoid compartment back to the stroma through the ATP synthase complex, whose catalytic heads use this proton-motive force to synthesise ATP on the stroma side of the membrane, where it is used to help drive the sugar synthesis during the Calvin cycle (light independent reactions).

1961: DNA found in plastids (*Ris* and *Plaut*)

1962-65: *Mortenson* described a nonheme-iron-containing protein they isolated from *Clostridium pasteurianum* and other bacteria which was later found to function as an electron carrier and named it ferredoxin. A connection between ferredoxin and photosynthesis was established when *Tagawa* and *Arnon* crystallised it and found it to mediate the photoreduction of $NADP^+$ by spinach chloroplasts.

Nonetheless, in 1963 *Shin* and co-workers demonstrated that ferredoxin could not react directly with $NADP^+$, and *Shin* and *Arnon* confirmed in 1965 that a photochemically reduced ferredoxin transferred electrons to a flavoprotein enzyme called ferredoxin-$NADP^+$ reductase. Thus, the first possible sequence of events leading from photo-oxidised chlorophyll via electron carriers to NADPH had been unravelled.

1965/66: *Kortschak et al.* and *Hatch & Slack* described a very efficient CO_2 binding mechanism which is used in some tropical grasses before the Calvin cycle (C_4-plants).

1965: *Pressmann* identified ionophores in membranes.

1965: The ultrastructure of chloroplasts was revealed by *A. Frey-Wyssling* and *K. Muhlethaler* and their structure was found to agree well with physiological data and hypotheses.

1965: *Arnon* and *Crane* identified seven benzoquinones as normal constituents of chloroplasts (plastoquinones A, B, C and D and α-, β-, and γ-tocopherolquinone.

1970: *David Walker* was able to demonstrate in isolated 'intact' chloroplasts the obligatory coupling of photosynthetic CO_2 uptake and O_2 evolution and later he also confirmed the required movement of sugars (and phosphate) in and out of the chloroplast for CO_2 fixation to occur.

1970: The theory of endosymbiosis was first formulated by *Lynn Margulis.*

1971: *Vernon* summarised the current concept of the pigment assemblies in thylakoids including accessory pigments and reaction centre chlorophyll molecules.

1972: The *"Fluid-Mosaic-Model"* by *Singer* and *Nicholson.*

1988: The most notable achievement in the photosynthesis field in recent years was the crystallisation of the reaction centre from a purple photosynthetic bacterium and the determination of its structure by X-ray crystallographic analysis by *Deisenhofer, Huber* and *Michel* in Germany for which they were awarded the Nobel Prize in 1988.

Crystallisation of the light-harvesting complex and photosynthetic reaction centres from selected plants and cyanobacteria have been achieved. High-resolution X-ray analysis of these crystals in the dark and light phases, which is in progress, would help to unravel the structure-function relationships of the reaction centre components and to understand the molecular processes associated with O_2 evolution.

Understanding Photosynthesis

Sunlight plays a much larger role in our sustenance than we may expect; all the food we eat and all the fossil fuel we use is a product of photosynthesis, which is the process that converts energy in sunlight to chemical forms of energy that can be used by biological systems. Photosynthesis is carried out by many different organisms, ranging from plants to bacteria. The best known form of photosynthesis is the one carried out by higher plants and algae, as well as by cyanobacteria and their relatives, which are responsible for a major part of photosynthesis in oceans.

All these organisms convert CO_2 (carbon dioxide) to organic material by reducing this gas to carbohydrates in a rather complex set of reactions. Electrons for this reduction reaction ultimately come from water, which is then converted to oxygen and protons. Energy for this process is provided by light, which is absorbed by pigments (primarily chlorophylls and carotenoids). Chlorophylls absorb blue and red light and carotenoids absorb blue-green light but green and yellow light are not effectively absorbed by photosynthetic pigments in plants; therefore, light of these colours is either reflected by leaves or passes through the leaves. This is why plants are green.

Figure: Examples of photosynthetic organisms: leaves from higher plants flanked by colonies of photosynthetic purple bacteria (left) and cyanobacteria (right).

Other photosynthetic organisms, such as cyanobacteria (formerly known as blue-green algae) and red algae, have additional pigments called phycobilins that are red or blue and that absorb the colours of visible light that are not effectively absorbed by chlorophyll and carotenoids. Yet other organisms, such as the purple and green bacteria (which, by the way, look fairly brown under many growth conditions), contain bacteriochlorophyll that absorbs in the infra-red, in addition to in the blue part of the spectrum. These bacteria do not evolve oxygen, but perform photosynthesis under anaerobic (oxygenless) conditions.

These bacteria efficiently use infra-red light for photosynthesis. Infra-red is light with wavelengths above 700 nm that cannot be seen by the human eye; some bacterial species can use infra-red light with wavelengths of up to 1000 nm. However, most pigments are not very effective in absorbing ultraviolet light (<400 nm), which also cannot be seen by the human eye.

Light with wavelengths below 330 nm becomes increasingly damaging to cells, but virtually all light at these short wavelengths is filtered out by the atmosphere (most prominently the ozone layer) before reaching the earth. Even though most plants are capable of producing compounds that absorb ultraviolet light, an increased exposure to light around 300 nm has detrimental effects on plant productivity.

Reaction Centres and Antennae

Photosynthetic pigments come in a huge variety: there are many different types of (bacterio)chlorophyll, carotenoids, and phycobilins, differing from each other in their precise chemical structure. Pigments generally are bound to proteins, which provide the pigment molecules with the appropriate orientation and positioning with respect to each other.

Light energy is absorbed by individual pigments, but is not used immediately by these pigments for energy conversion. Instead, the light energy is transferred to chlorophylls that are in a special protein environment where the actual energy conversion event occurs:

> The light energy is used to transfer an electron to a neighbouring pigment. Pigments and protein involved with this actual primary electron transfer event together are called the reaction centre. A large number of pigment molecules (100-5000), collectively referred to as antenna, "harvest" light and transfer the light energy to the same reaction centre. The purpose is to maintain a high rate of electron transfer in the reaction centre, even at lower light intensities.

Many antenna pigments transfer their light energy to a single reaction centre by having this energy "hop" to another antenna pigment, and yet to another, etc., until the energy is "trapped" in the reaction centre. Each step of this energy transfer must be very efficient to avoid a large loss in the overall transfer process, and the association of the various pigments with proteins ensures that transfer efficiencies are high by having appropriate pigments close to each other, and by having an appropriate molecular geometry of the pigments with respect to each other.

An exception to the rule of protein-bound pigments are green bacteria with very large antenna systems: a large part of these antenna systems consists of a "bag" (named chlorosome) of up to several thousand bacteriochlorophyll molecules that interact with each other and that are not in direct contact with protein.

In many systems the size of the photosynthetic antenna is flexible, and photosynthetic organisms growing at low light (in the shade, for example) generally will have a larger number of antenna pigments per reaction centre than those growing at higher light intensity. However, at high light intensities (for example, in full sunlight) the amount of light that is absorbed by plants exceeds the capacity of electron transfer initiated by reaction centres.

Plants have developed means to convert some of the absorbed light energy to heat rather than to use the absorbed light necessarily for photosynthesis. However, in particular the first part of photosynthetic electron transfer in plants is rather sensitive to overly high rates of electron transfer, and part of the photosynthetic electron transport chain may be shut down when the light intensity is too high; this phenomenon is known as photoinhibition.

Photosynthetic Electron Transfer

The initial electron transfer (charge separation) reaction in the photosynthetic reaction centre sets into motion a long series of redox (reduction-oxidation) reactions, passing the electron along a chain of co-factors and filling up the "electron hole" on the chlorophyll, much like in a bucket brigade.

All photosynthetic organisms that produce oxygen have two types of reaction centres, named photosystem II and photosystem I (PS II and PS I, for short), both of which are pigment/protein complexes that are located in specialised membranes called thylakoids.

In eukaryotes (plants and algae), these thylakoids are located in chloroplasts (organelles in plant cells) and often are found in membrane stacks (grana).

Prokaryotes (bacteria) do not have chloroplasts or other organelles, and photosynthetic pigment-protein complexes either are in the membrane around the cytoplasm or in invaginations thereof (as is found, for example, in purple bacteria), or are in thylakoid membranes that form much more complex structures within the cell (as is the case for most cyanobacteria).

Figure: Artist's rendition of a leaf (bottom), thylakoids within a chloroplast (middle), and a photosystem in thylakoids (top).

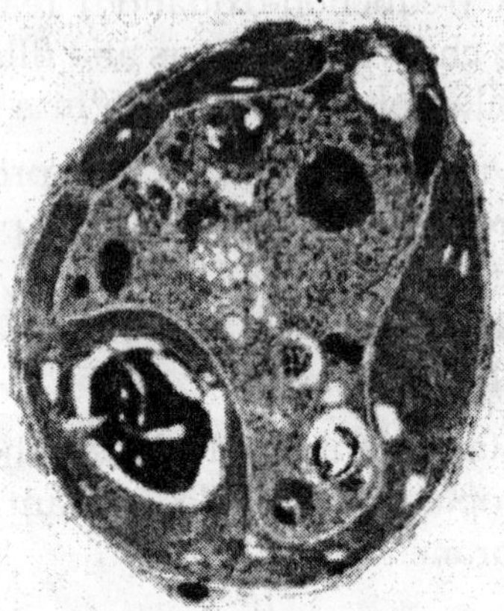

Figure: Electron micrograph of a thin section of an algal cell. The cup-shaped structure around the edge of the cell (open near the top) is the chloroplast. The structures resembling mostly parallel lines in the chloroplast are the thylakoid membranes.

Figure: Freeze-fracture electron micrograph of a cyanobacterial cell, showing exposed thylakoid membrane surfaces (upper right). Thylakoids are stacked like folded pancakes, and this image represents a surface-cut through these thylakoids.

All chlorophyll in oxygenic organisms is located in thylakoids, and is associated with PS II, PS I, or with antenna proteins feeding energy into these photosystems. PS II is the complex where water splitting and oxygen evolution occurs. Upon oxidation of the reaction centre chlorophyll in PS II, an electron is pulled from a nearby amino acid (tyrosine) which is part of the surrounding protein, which in turn gets an electron from the water-splitting complex.

From the PS II reaction centre, electrons flow to free electron carrying molecules (plastoquinone) in the thylakoid membrane, and from there to another membrane-protein complex, the cytochrome $b_6 f$ complex. The other photosystem, PS I, also catalyses light-induced charge separation in a fashion basically similar to PS II: light is harvested by an antenna, and light energy is transferred to a reaction centre chlorophyll, where light-induced charge separation is initiated.

However, in PS I electrons are transferred eventually to NADP (nicotinamid adenosine dinucleotide phosphate), the reduced form of which can be used for carbon fixation. The oxidised reaction centre chlorophyll eventually receives another electron from the cytochrome $b_6 f$ complex. Therefore, electron transfer through PS II and PS I results in water oxidation (producing oxygen) and NADP reduction, with the energy for this process provided by light (2 quanta for each electron transported through the whole chain).

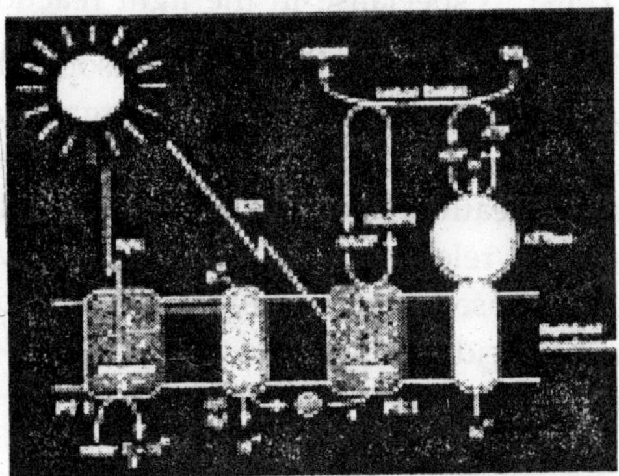

Figure: Overview of photosynthetic processes as they occur in plants, algae, and cyanobacteria.

Carbon Fixation

Electron flow from water to NADP requires light and is coupled to generation of a proton gradient across the thylakoid membrane. This proton gradient is used for synthesis of ATP (adenosine triphosphate), a high-energy molecule. ATP and reduced NADP that resulted from the light reactions are used for CO_2 fixation in a process that is independent of light. CO_2 fixation involves a number of reactions that is referred to as the Calvin-Benson cycle. The initial CO_2 fixation reaction involves the enzyme ribulose-1,5-bisphosphate carboxylase/ oxygenase (Rubisco), which can react with either oxygen (leading to a process named photorespiration and not resulting in carbon fixation) or with CO_2. The probability with which Rubisco reacts with oxygen depends on the relative concentrations of the two compounds at the site of the reaction.

In all organisms CO_2 is by far the preferred substrate, but as the CO_2 concentration is very much lower than the oxygen concentration, photorespiration does occur at significant levels. To boost the local CO_2 concentration and to minimise the oxygen tension, some plants (referred to as C_4 plants) have set aside some cells within a leaf (named bundle-sheath cells) to be involved primarily in CO_2 fixation, and others (named mesophyll cells) to specialise in the light reactions:

> ATP, CO_2 and reduced NADP in mesophyll cells is used for synthesis of 4-carbon organic acids (such as malate), which are transported to bundle sheath cells. Here the organic acids are converted releasing CO_2 and reduced NADP, which are used for carbon fixation. The resulting 3-carbon acid is returned to the mesophyll cells. The bundle sheath cells generally do not have PS II activity, in order to minimise the local oxygen concentration. However, they retain PS I, presumably to aid in ATP synthesis. Even though C-4 plants have reduced amounts of

photorespiration, the amount of ATP they need per amount of CO_2 fixed is a little higher than in other plants, and therefore their total production rate is similar to that of plants with higher rates of photorespiration.

Some plants living in desert climates, such as cacti, keep their stomates closed during the day to minimise evaporation (stomates are openings in the leaf surface to enhance gas exchange). These plants take up CO_2 during the night when the stomates are open, and temporarily bind the CO_2 to organic acids in the leaf. During the day the CO_2 is released from the acids and used for photosynthesis. Plants using this mechanism of CO_2 fixation are called CAM (Crassulacean Acid Metabolism) plants.

Figure: The light reactions of photosynthesis stop when the sun goes down. However, CO_2 fixation can continue as long as ATP and NADPH is available. In cacti and other succulents CO_2 uptake by the plant occurs primarily at night.

Hatch-Slack Photosynthetic Pathway

In the 1960s, two Australian scientists, M.D. Hatch and C.R. Slack described a new pathway for carbon fixation in photosynthesis. The Hatch-Slack pathway describes a biochemical system in which carbon is first incorporated into the four-carbon molecule, oxaloacetic acid. Earlier, in the 1950s, an American scientist, Melvin Calvin, uncovered a chemical pathway for photosynthetic carbon fixation that came to be called the Calvin Cycle. In his scheme, carbon was first fixed into the three carbon compound, 3-phosphoglyceric acid (PGA). He was awarded the Nobel Prize for this discovery in 1961. For a time the Calvin Cycle was thought to be the only carbon fixation pathway. Hatch and Slack's surprising discovery proved otherwise. Their efforts were aided by the work of G.O.

Burr and his colleagues at the Sugar Cane Research Institute in Hawaii who reported that PGA was not the first photosynthetic product in sugarcane plants, a species that is especially efficient in photosynthesis. The experimental procedure for discovery of early products involved exposing photosynthesising plant material for a very brief time to carbon dioxide enriched with radioactive 14-C. The plant cells were then extracted and analysed by chromatography to determine what chemicals had incorporated radioactivity carbon from the carbon dioxide. Burr's laboratory found that PGA was not the first compound to become radioactive, but it remained for Hatch and Slack to sort out details of the newly-discovered pathway.

The two fixation mechanisms are sometimes called the C_3 and C_4 pathways based on the number of carbons in the initial product. Each pathway has its own unique catalysing enzyme and acceptor molecule. The molecule that accepts the carbon in the C_4 pathway is a three-carbon compound called phosphoenolpyruvate (PEP) and the catalysing enzyme is PEP carboxylase. The acceptor molecule for the C_3 pathway is a five-carbon sugar called Ribulose 1,5 bisphosphate (RuBP), which splits into 2 molecules of PGA after adding a carbon, and the enzyme is RuBP carboxylase. The two carboxylating enzymes use different forms of carbon dioxide as substrate. RuBP carboxylase incorporates carbon dioxide, whereas PEP carboxylase uses the hydrated bicarbonate ion form. The enzymes also differ in their cellular location. RuBP carboxylase is located in the chloroplast, and PEP carboxylase in the cytoplasm. Sugarcane, and other plants that first fix carbon by the Hatch-Slack pathway also use the Calvin cycle in a curious system that involves the uptake of carbon dioxide twice. The complex double fixation process is commonly found in tropical plants and other plants that are highly efficient in photosynthesis. This efficiency may result from PEP carboxylase's greater affinity for carbon dioxide and its ability to assist the less efficient enzyme.

Kranz Anatomy

The leaves of green plants using the C_4 or Hatch-Slack pathway for photosynthetic carbon fixation almost invariably have a specialised internal arrangement of cells surrounding the vascular bundles that is called Kranz anatomy. Kranz, the German word for "halo," or "wreath,"refers to a ring of mesophyll cells just to the outside of another ring of large bundle-sheath cells, both of which encircle the vascular bundle. In transverse sections viewed under the microscope, the two cell layers give the appearance of a wreath surrounding each bundle. In addition to the unique "wreaths," other features that typify leaves with Kranz anatomy include small intercellular spaces, and frequent veins.

Kranz anatomy and the C_4 photosynthetic pathway are especially characteristic of tropical grasses such as sugar cane, where it was first discovered, and corn, although it has also been found in other plants. It is sometimes possible to distinguish C_4 plants by their dark green veins, which are a consequence of the chlorophyll rich "wreaths" surrounding the conductive tissue. The chloroplasts of mesophyll cells typically contain chlorophyll-bearing internal membranous structures called grana, while the large and conspicuous chloroplasts in bundle-sheath cells lack grana or have only a poorly developed type. During active photosynthesis, bundle-sheath chloroplasts tend to form larger and more numerous starch grains than the mesophyll chloroplasts.

Plants with Kranz anatomy and the C_4 photosynthetic pathway tend to be highly efficient in photosynthesis. They generally have higher maximum rates of photosynthesis, and become light saturated at higher light intensities enabling them to capture and store large amounts of light energy even in tropical areas. They are able to photosynthesise more effectively at higher temperatures, and at low carbon dioxide concentrations that severely limit photosynthesis in less efficient C_3 plants. The basis for the superior photosynthetic ability of C_4 plants with Kranz anatomy is based in part on cooperation between the C_3 and C_4 photosynthetic pathways, both of which occur in these plants, and a peculiar system that involves the double fixation of carbon dioxide. Carbon dioxide is first fixed in the mesophyll cells by the C_4 pathway with a three-carbon compound, phosphoenolpyruvate, as the acceptor molecule, and the four-carbon molecule, oxaloacetate, as the product. Oxaloacetate is quickly converted to malate using reducing power produced in mesophyll chloroplasts. The malate is transported to bundle-sheath cells where it releases carbon dioxide. The released carbon dioxide is quickly captured by the C_3 pathway, forming a three-carbon compound, 3-phosphoglycerate, and is then incorporated into sugars and starch.

Increasing CO_2 Levels

The amount of overall CO_2 fixation in plants growing under optimal conditions is limited primarily by the amount of CO_2 available. Therefore, the increase of CO_2 in the atmosphere will lead to somewhat higher rates of plant growth in environments where the CO_2 concentration limits growth rates. This is usually the case in an agricultural setting, where nutrients and water availability are not limiting. However, also in natural conditions, where limitations other than the CO_2 concentration will generally limit plant productivity, plant productivity has been found to often increase upon increasing the CO_2 concentration.

Photosynthesis and Respiration

Respiration in plants, as in all living organisms, is essential to provide metabolic energy and carbon skeletons for growth and maintenance. As such, respiration is an essential component of a plant's carbon budget. Depending on species and environmental conditions, it consumes 25-75 per cent of all the carbohydrates produced in photosynthesis even more at extremely slow growth rates. Respiration in plants can also proceed in a manner that produces neither metabolic energy nor carbon skeletons, but heat.

This type of respiration involves the cyanide-resistant, alternative oxidase; it is unique to plants, and resides in the mitochondria. The activity of this alternative pathway can be measured based on a difference in fractionation of oxygen isotopes between the cytochrome and the alternative oxidase. Heat production is important in some flowers to attract pollinators; however, the alternative oxidase also plays a major role in leaves and roots of most plants. A common thread throughout this volume is to link respiration, including alternative oxidase activity, to plant functioning in different environments.

Virtually all oxygen in the atmosphere is thought to have been generated through the process of photosynthesis. Obviously, all respiring organisms (including plants) utilise this oxygen and produce CO_2. Thus, photosynthesis and respiration are interlinked, with each process depending on the products of the other. The global amount of photosynthesis is on the order of a trillion kg of dry organic matter produced per day, and respiratory processes convert about the same amount of organic matter to CO_2. A large part (probably the majority) of photosynthetic productivity occurs in open oceans, mostly by oxygenic prokaryotes. Without photosynthesis, the oxygen in the atmosphere would be depleted within several thousand years. It should be emphasised that plants respire just like any other higher organism, and that during the day this respiration is masked by a higher rate of photosynthesis.

Diversity of Photosynthetic Organisms

Even though plants are the most visible representatives of photosynthetic organisms, it should be emphasised that many other types of photosynthetic organisms exist. All photosynthetic bacteria other than the cyanobacteria and their relatives use only one photosystem, and for thermodynamic reasons they cannot use water as the ultimate electron donor. Instead, they can use reduced compounds such as H_2S as donor. However, CO_2 fixation occurs in these organisms. Some of these photosynthetic bacteria appear to have retained an evolutionary ancient arrangement of their photosynthetic apparatus, and are of interest for the analysis of evolutionary relationships of photosynthetic systems.

An extensive group of these photosynthetic bacteria, the heliobacteria, was discovered rather recently in the 1980s. The first representative of this group was isolated by the group of Dr. Howard Gest from a soil sample collected on the campus of Indiana University, and this isolation was the result of a fortunate coincidence of serendipitous events. Analysis of the heliobacterial reaction centre has helped to lay the basis for the current concept that all photosynthetic reaction centres from the large variety of photosynthetic organisms are related to each other. The majority of bacteria cannot be maintained in pure culture (that is, without other organisms).

This has essentially limited analysis of photosynthetic prokaryotes to the relatively small group of organisms that can be grown in pure culture. It is likely that the actual diversity of photosynthetic organisms is much larger than is known thus far. Indeed, species with novel photosynthetic properties are reported virtually every year. For example, recently an organism was reported that has chlorophyll d (a chlorophyll that is very rare in nature) as the main pigment. Moreover, several years ago, previously undetected and very small chlorophyll *a/b*-containing prokaryotes were recognised to be the major contributors to photosynthetic production in the open ocean. This emphasises that much relating to biodiversity and photosynthesis is still to be discovered, and that these discoveries are not limited to tropical rainforests and other ecological settings of large popular interest.

Evolution

In eukaryotes, photosynthesis takes place in the chloroplast, which has long been known to have prokaryotic features. Chloroplasts are thought to have evolved from a cyanobacterium (or close relative) that was in a symbiotic relationship with a eukaryotic, non-photosynthetic cell and was engulfed inside this cell. The cyanobacterium and the eukaryotic cell presumably were in a mutually beneficial relationship (endosymbiosis), with the photosynthetic organism sharing some of its produced carbohydrates with the host, and the host providing the photosynthetic bacterium with other compounds. The prokaryote slowly gave up its independence as well as its cell wall, and some of its genetic information was transferred to the nucleus of its eukaryotic host.

The resulting chloroplast maintains a small, prokaryote-like circular DNA of its own (DNA is material carrying genetic information); this DNA contains the genetic blueprint to make many of the membrane proteins needed in the chloroplast, which apparently are not easily targeted to and/or transported into the chloroplast. Occasionally, photosynthetic organisms are found where the chloroplast has retained a little more of the original cyanobacterial features. For example, in algae such as Cyanophora paradoxa plastids (called cyanelles) are found that resemble cyanobacteria in their overall morphology as well as in the fact that they are surrounded by a cell wall.

Not all chloroplasts have resulted from a single endosymbiotic event, but apparently from multiple events that occurred independently. Chloroplasts from higher plants and many green algae probably all result from the same endosymbiotic event, whereas chloroplasts from red and brown algae and from diatoms are the result of one or more other events.

The situation is even more complicated in cryptomonads, a type of algae, and chlorachniophytes, photosynthetic amoebae, which apparently are the result of an endosymbiotic event of a eukaryotic alga in a eukaryotic host. The nucleus of the endosymbiont has been mostly degraded, resulting in a chloroplast enveloped by four membranes.

Early Events

Chlorophyll is used by all photosynthetic organisms as the link between excitation energy transfer and electron transfer. Of particular note is the rate with which these transfer reactions need to occur. As the lifetime of the excited state is only several nanoseconds (1 nanosecond (ns) is 10^{-9} s), after absorption of a quantum, energy transfer and charge separation in the reaction centre must have occurred within this time period.

Energy transfer rates between pigments are very rapid, and charge separation in reaction centres occurs in 3-30 picoseconds (1 picosecond (ps) is 10^{-12} s). Subsequent electron transfer steps are significantly slower (200 ps - 20 ms) but, nonetheless, the electron transport chain is sufficiently fast that at least a significant part of the absorbed sunlight can be used for photosynthesis.

However, in the presence of excess light, damage may occur, which may originate from the formation of chlorophyll in "triplet state". In a triplet state two electrons in the outer shell have identical rather than opposite spin orientation. This triplet chlorophyll readily reacts with oxygen, leading to the very reactive singlet oxygen, which can damage proteins. To counter this damaging reaction, carotenoids are usually present in close vicinity to chlorophylls. Many carotenoids efficiently "quench" triplet states of chlorophyll, thus avoiding formation of singlet oxygen. Chlorophyll in its free form is very toxic in the light in the presence of oxygen, because a close interaction with carotenoids is not always available under such circumstances. Therefore, all chlorophyll in a cell in aerobic organisms is bound to proteins, generally with carotenoids bound to the same protein.

Structure Determinations

Because of the strict requirements of positioning of pigments and electron transfer intermediates to allow efficient electron transfer and minimal damage, the structure of pigment-protein complexes involved in photosynthesis is critical. With the exception of specific antenna complexes (such as phycobilisomes in cyanobacteria and chlorosomes in green bacteria), pigment-binding proteins are usually hydrophobic membrane proteins.

This initially hampered attempts to elucidate the structure of these complexes, as membrane proteins do not readily form the well-ordered crystals that are needed for high-resolution X-ray diffraction studies. However, in the 1980s the first structure of a membrane protein complex, the photosynthetic reaction centre from a purple bacterium, was determined at high resolution (about 3 Å; in comparison, the distance between neighbouring atoms in a molecule is about 1 Å).

Investigators from the Max Planck Institute in Martinsried (Germany) who were involved with this work, most notably Hartmut Michel and Johann Deisenhofer, received a Nobel Prize in chemistry for this research. Since then, the structure of various reaction centres and antenna complexes has been determined at resolutions between 2 and 4 Å. Figure presents the structure of the photosynthetic reaction centre from the purple bacterium *Rhodobacter sphaeroides*.

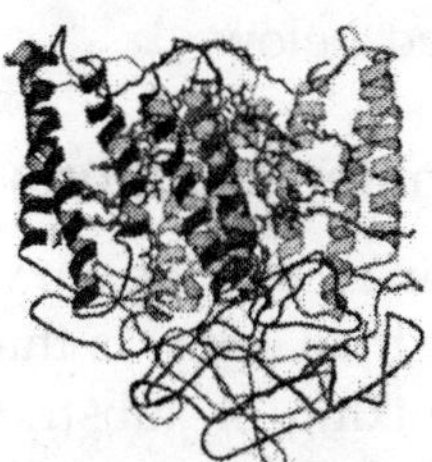

Figure: Molecular structure of a bacterial reaction centre. Co-factors are indicated in red. The three proteins making up the reaction centre are in blue, yellow, and green. Ribbons in the protein represent helices through the membrane. Wire-like protein regions represent domains outside of the membrane.

Similarities between Reaction Centres

Surprisingly, structural comparison of reaction centres from different photosynthetic systems showed that these reaction centres are basically similar to each other in terms of their overall three-dimensional structure. The basic reaction centre unit consists of a protein complex with 10 transmembrane helices originating from either two identical protein subunits or from two similar polypeptides of common evolutionary origin. Each of these polypeptides contribute five membrane-spanning helices, and bind 2-3 chlorophylls (or, in the case of anoxygenic bacteria, bacteriochlorophylls). The fourth membrane spans from each subunit are held together by two chlorophylls, which are the chlorophylls in the reaction centre that can be oxidised (give up an electron) upon excitation. Directly associated with the reaction centre are proteins that bind antenna pigments. In the case of PS I and similarly organised reaction centres from green and heliobacteria, the antenna portion, with six transmembrane helices, is attached to the N-terminal end of the reaction centre proteins.

Implications of Photosynthesis Studies

Photosynthesis has been studied in significant detail and photosynthetic systems are used frequently for development and application of advanced technologies, because photosynthetic systems are fairly well understood, are complex, and often undergo rather unusual bio-chemical reactions. Some examples are provided below.

Rapid Electron Transfer Reactions

A major difficulty in measuring enzyme kinetics at relatively short time scales (less than 1 ms) is that "traditional" enzyme reactions require a mixing of substrate and enzyme, which usually takes a relatively long time. Kinetic analysis of light-driven reactions such as photosynthetic electron transport have a great advantage in this respect: reactions can be triggered simply by a light pulse, which can be even shorter than 1 ps.

Moreover, many of the components participating in electron transfer have different absorption spectra depending on whether they are in the oxidised or reduced form. Using laser spectroscopy methods or more standard optical spectroscopy, it is relatively simple to follow the electron around on a timescale between 1 ps and several ms. The primary charge separation occurs in several ps, and reactions become gradually slower as they involve components that are further away from the reaction centre. Because of the fast speed of early reactions, the electron and the "electron hole" are physically separated rapidly by a large distance (the electron generally has travelled about 2 nm to the other side of the membrane within 1 ns after charge separation), so that back reactions (charge recombinations) are not favourable anymore.

Unpaired electrons on reactants that are transiently formed during redox reactions involving transfer of a single electron in many instances can be detected using electron paramagnetic resonance (EPR) and derived techniques (including ENDOR, electron nuclear double resonance, and ESEEM, electron spin echo envelope modulation). Many of these techniques can be used to kinetically follow redox reactions, and provide detailed information regarding electron spin distributions, etc. Therefore, photosynthetic membranes and reaction centres have a prominent place as experimental systems in bio-chemistry and bio-physics.

Organic Molecules Mimicking Reaction Centres

Reaction centres are essentially an assembly of co-factors, held in the appropriate position and orientation by the protein environment. Several groups have used the natural system as a model to design organic molecules where the equivalents of the different co-factors are linked together by covalent bonds of various lengths. The result is the creation of a number of sophisticated molecules that serve as "artificial reaction centres". The more advanced molecules consist of two chlorophyll-type molecules linked together (one serves as the electron donor, the other as the acceptor), with the electron-accepting molecule linked to two quinones, which serve as electron acceptors in the natural system. The electron donating chlorophyll analog is linked covalently to a carotenoid, which can donate an electron to the oxidised chlorophyll. Upon excitation of the chlorophyll, a charge separation occurs resulting in an oxidised carotenoid and a reduced quinone. This charge-separated state is formed with high efficiency. An example of such a molecule is presented in Figure.

Figure: Molecular model of an artificial reaction centre. The two bulky structures in the middle are chlorophyll-like components, flanked by a carotenoid (left) and quinones (right).

Such molecules can be introduced into liposomes (artificial membrane vesicles) in a specific orientation, and when these are excited by light, a charge separation will occur across the liposome membrane. This results in an electric potential or proton gradient across the liposome membrane, which may be used for a variety of purposes, including ATP synthesis (the latter requires introduction of the ATP synthesising enzyme into the liposome membrane). The groups of Ana and Tom Moore and Devens Gust at Arizona State University are leaders in developments in this area.

Genetic Modification and Protein Engineering

Because of the ease of detailed functional analysis of reactions and their rates in photosynthetic systems, reaction centre complexes are frequently used to determine the consequences of small alterations in the polypeptides on the functional characteristics of the co-factors. Changes at single amino acid residues in the reaction centre complex are sufficient to introduce large changes in the properties of co-factors, which in turn leads to altered electron transfer rates and efficiencies. Single amino acid changes at specific sites in the protein are easily introduced by genetic modification techniques, and resulting functional changes can be studied. An elegant example of such an approach is the modification of the midpoint redox potential of the bacteriochlorophyll in the reaction centre of purple bacteria. The midpoint redox potential is correlated with the ease with which an electron is given off after excitation and is taken up by the oxidised bacteriochlorophyll. Jim Allen, Joann Williams and co-workers at Arizona State University found that creating or deleting hydrogen bonds between the protein and the bacteriochlorophyll changed the mid-point redox potential of this bacteriochlorophyll in a rather predictable manner.

In this way, reaction centre complexes can be built with different oxidising strengths, and effects on reaction rates and ultimately the effectiveness of alternate electron donors can be determined. Mutational analysis of photosynthesis proteins is simple in several bacterial systems. The reasons why this is so in selected cyanobacteria and purple and green bacteria are that (1) foreign DNA is taken up by the cell spontaneously or is introduced easily by other means such as electroporation ("electric shock"), (2) once the DNA is inside it is incorporated into the organism's genome at one predictable and specific site by means of a process named homologous double recombination, and (3) the organism can be propagated without relying on photosynthesis, for example using an added carbohydrate source.

Genetic approaches involving directed mutagenesis have proven to be very useful in studying photosynthetic electron transfer and will be of increasing relevance for the design of photosynthetic organisms for biotechnological uses. By this method the function of a large number of genes has been probed, and the role of individual domains and residues has been determined. Genomic sequencing projects are very useful in this respect, and the complete DNA sequence of one photosynthetic organism is already known. From the DNA sequence, the potential of the organism can be determined. The entire 3,573,470 nucleotide-long genomic DNA sequence of the transformable cyanobacterium *Synechocystis* 6803 was determined by Satoshi Tabata and co-workers at the Kazusa DNA Research Institute in Japan.

This organism is used by several researchers, including in the Vermaas group at Arizona State University, to elucidate the role of many proteins thought to be involved in photosynthetic or other physiological processes. Meanwhile, other groups are working on the genomic sequence of two purple bacteria. With the genomic sequence in hand, the role of specific genes can be found by amplifying the gene of interest by means of polymerase chain reaction, cloning it into a plasmid, replacing the gene by a selectable marker (i.e., a piece of DNA coding for a protein inactivating a particular antibiotic, thus conferring antibiotic resistance), and analysing the functional characteristics of resulting mutants.

Genetic Modification of Higher Plants

Globally, improving plant productivity by genetic means has been an important goal for many years. Even though initially it was hoped that crop productivity could be boosted by "improving" the photosynthesis process, it has become increasingly apparent that improved productivity will be best accomplished by genetic modification of crop plants to introduce insect or pathogen resistance, or to yield improved vitality under marginal conditions (for example, at high salinity, which has become a significant factor in many agricultural areas because of continued irrigation). A pertinent example is the development of varieties of cotton and other crops that express a pro-toxin from a bacterium, Bacillus thuringiensis, that is converted to a toxin in the midgut of particular insects such as caterpillars but not in other life forms. This allows efficient biological control as long as the insects do not develop resistance to this compound.

Genetic modifications to intrinsically and significantly improve the photosynthesis process have not yet been successful. The reason for the apparent inability to "improve" the photosynthesis process itself presumably is related to the fact that photosynthetic systems have evolved over a relatively long period of time, and that the selection factors have not changed significantly in recent history. This has led to the emergence of a very effective photosynthetic apparatus that is difficult to improve upon by simply changing some amino acid residues or by introducing or deleting some genes. If relatively simple changes could have significantly improved the photosynthesis process per se, Mother Nature would have already found these as natural mutations are rather frequent.

However, it is possible that significant progress in this area can made in the future, if new design paradigms for enzyme function and specificity can be developed. For example, if protein structures (particularly the structure of the active site) can be better modelled and predicted, one should be able to further improve upon the Rubisco specificity of CO_2 vs. oxygen. It is also important in this respect to determine what the rate-limiting step in the process is under natural conditions. Even though more effective light capture by crop plants might be considered by introduction of antenna pigments that absorb in the green and yellow region of the light spectrum, light capture and the light reactions usually are not limiting plant productivity in agricultural settings. Therefore, such modifications in a plant will result in increased productivity only in light-limiting settings.

Future Directions

Light energy is cheap, clean, and essentially inexhaustible. With limited supplies of fossil fuel and increasing concern about CO_2 emissions, further development of technologies that make use of solar energy is inevitable. Current silicon-based technologies for the harvesting of solar energy require a very energy-intensive production process and even though they have improved significantly over the years in their efficiency, further development of photosynthesis-based technologies for energy collection is certainly warranted. However, photosynthesis and related processes can be applied to many more areas than just solar energy conversion. Realising that novel designs and applications of light-mediated processes have enormous promise in the next decade and beyond, Arizona State University has started an initiative, Project Ingenhousz, named after the 18th century physician who discovered that light is needed for oxygen evolution by plants, and that only green parts of the plant carry out this process.

The goal of the initiative is to capitalise on research progress and ideas in this area, and to more effectively interface academia, where many of the discoveries are made, with the private sector, where such discoveries are worked out further and applied.

For example, there are a myriad of possible applications of artificial reaction centres and related molecules in nanotechnology. Many synthetic pigments also have found biomedical uses in tumour detection, as they -for unknown reasons- tend to accumulate preferentially in tumors and are highly fluorescent and thus easily detectable in a patient whom is being operated on to surgically remove a tumour.

In the biotechnology field, photosynthetic organisms are likely to play an increasing role in (over)production of enzymes, pharmaceuticals, nutraceuticals, etc., which until now are produced primarily by genetically modified heterotrophic microorganisms such as yeast and selected bacteria. A major advantage of photosynthetic organisms is that no fixed-carbon source needs to be added for growth and, therefore, production costs are lower and the chances of contamination with other microorganisms are less.

There are several ways to modify organisms to have them (over)produce useful compounds. One is to introduce specific genes under a strong promoter, leading to high expression of these genes and to synthesis of a "new" enzyme. Another way is to delete genes so that substrates will accumulate. For this to be optimally successful the metabolic pathways of an organism need to be fairly well understood, and genomic DNA sequences are an important step in this direction. A third way to produce a new compound is to utilise an existing enzyme and to modify the site where the substance to be converted (the substrate) binds, so that a different substrate can be bound and a different product can be formed. With any of these approaches, selection for randomly generated combinatorial mutants with specific properties also has been proven to be effective.

With an increasing arsenal of genomic sequences, and with improving knowledge regarding determinants of protein structure and co-factor binding, any of these three approaches are very promising.

Another potential application of photosynthetic organisms is in bioremediation. Bioremediation is the clean-up of environmental (soil or water) pollutants by biological means. An example is the biological breakdown of toxic organic compounds into innocuous products. Also, remediation of nitrate from drinking water supplies is becoming an increasingly pressing issue. The advantage of using photosynthetic organisms is that no external energy source needs to be provided for growth of the organism if it is in the light, making these organisms very suitable for remediation of aqueous surface environments.

Another utilisation of photosynthetic organisms is to have these organisms use solar energy to produce clean-burning fuels. Even under natural conditions some photosynthetic systems such as algae can produce hydrogen, which probably is the cleanest fuel as it reacts with oxygen to produce water. However, the cheapest and most universal electron donor of all, water, upon oxidation in PS II forms oxygen, which is not very safe in combination with hydrogen. For this reason methods have been sought to temporally separate oxygen evolution and hydrogen production in algae; in the laboratory this can be achieved by sulphur deprivation, which preferentially inhibits photosynthesis. Another option would be to use photosynthetic organisms for methane production. Even though methane upon combustion will form CO_2, the overall atmospheric CO_2 balance would not be disturbed as an equal amount of CO_2 will have been taken out of the atmosphere upon methane production by the photosynthetic organism..

Research in photosynthesis in all its facets has proven to have opened many doors in a variety of disciplines, ranging from biophysics to plant physiology. Progress has been driven by an interdisciplinary approach to this complex, yet fascinating, spectrum of problems, challenges, and opportunities. Photosynthesis is the basis of our food and energy supply, and innovative utilisation of solar energy is likely to be of increasingly critical importance in the future. This, together with novel uses of photosynthetic principles for other purposes, make it likely that photosynthesis and its applications will help to shape an increasingly broad area of exciting discoveries and innovative ideas. Chemicals would have been present in all plants, leading scientists to hypothesise that C_4 mechanisms evolved several times independently in response to a similar environmental condition, a type of evolution known as convergent evolution.

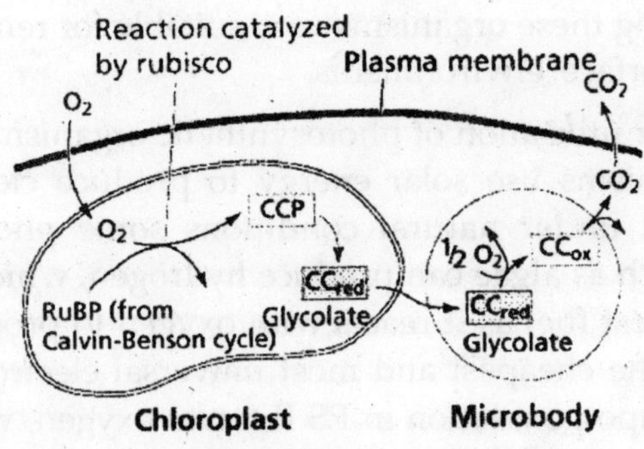

Photorespiration

We can see anatomical differences between C_3 and C_4 leaves.

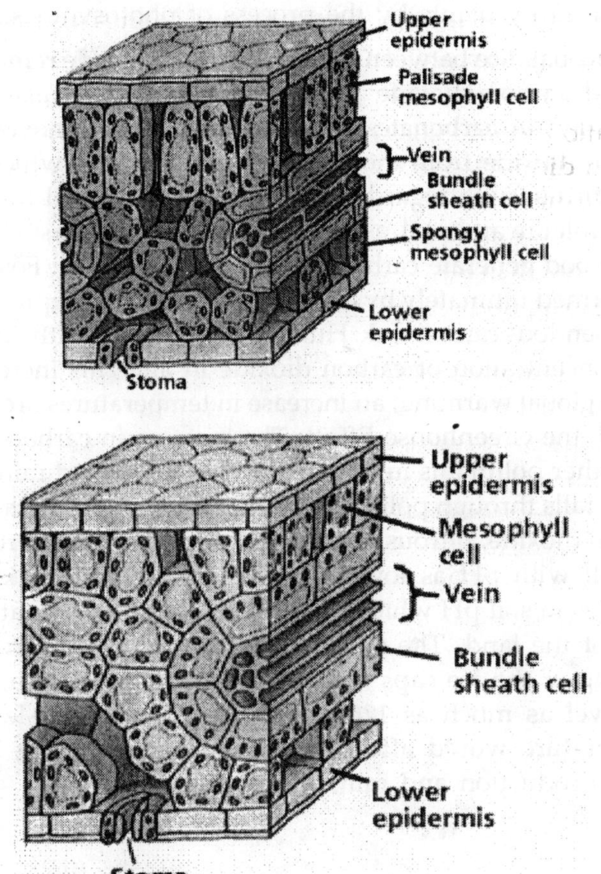

Leaf anatomy of a C_3 (top) and C_4 (bottom) plant.

Carbon Cycle

Plants may be viewed as carbon sinks, removing carbon dioxide from the atmosphere and oceans by fixing it into organic chemicals. Plants also produce some carbon dioxide by their respiration, but this is quickly used by photosynthesis. Plants also convert energy from light into chemical energy of C-C covalent bonds. Animals are carbon dioxide producers that derive their energy from carbohydrates and other chemicals produced by plants by the process of photosynthesis.

The balance between the plant carbon dioxide removal and animal carbon dioxide generation is equalised also by the formation of carbonates in the oceans. This removes excess carbon dioxide from the air and water (both of which are in equilibrium with regard to carbon dioxide). Fossil fuels, such as petroleum and coal, as well as more recent fuels such as peat and wood generate carbon dioxide when burned. Fossil fuels are formed ultimately by organic processes, and represent also a tremendous carbon sink. Human activity has greatly increased the concentration of carbon dioxide in air. This increase has led to global warming, an increase in temperatures around the world, the Greenhouse Effect. The increase in carbon dioxide and other pollutants in the air has also led to acid rain, where water falls through polluted air and chemically combines with carbon dioxide, nitrous oxides, and sulphur oxides, producing rainfall with pH as low as 4. This results in fish kills and changes in soil pH which can alter the natural vegetation and uses of the land. The Global Warming problem can lead to melting of the ice caps in Greenland and Antarctica, raising sea-level as much as 120 metres. Changes in sea-level and temperature would affect climate changes, altering belts of grain production and rainfall patterns.

Photosynthetic Organisms

All life can be divided into three domains, Archaea, Bacteria and Eucarya, which originated from a common ancestor (Woese *et al.,* 1990).

Historically, the term photosynthesis has been applied to organisms that depend on chlorophyll (or bacteriochlorophyll) for the conversion of light energy into chemical free energy (Gest, 1993)

These include organisms in the domains Bacteria (photosynthetic bacteria) and Eucarya (algae and higher plants).

The most primitive domain, Archaea, includes organisms known as halobacteria, that convert light energy into chemical free energy. However, the mechanism by which halobacteria convert light is fundamentally different from that of higher organisms because there is no oxidation/ reduction chemistry and halobacteria cannot use CO_2 as their carbon source.

Consequently some biologists do not consider halobacteria as photosynthetic (Gest 1993).

Oxygenic Photosynthetic Organisms

The photosynthetic process in all plants and algae as well as in certain types of photosynthetic bacteria involves the reduction of CO_2 to carbohydrate and removal of electrons from H_2O, which results in the release of O_2. In this process, known as oxygenic photosynthesis, water is oxidised by the photosystem II reaction centre, a multisubunit protein located in the photosynthetic membrane. Years of research have shown that the structure and function of photosystem II is similar in plants, algae and certain bacteria, so that knowledge gained in one species can be applied to others.

This homology is a common feature of proteins that perform the same reaction in different species. This homology at the molecular level is important because there are estimated to be 300,000-500,000 species of plants. If different species had evolved diverse mechanisms for oxidising water, research aimed at a general understanding of photosynthetic water oxidation would be hopeless.

Except for plants and some unicellular protists, which are secondary photosynthetic due to the presence of chloroplasts that have originated from cyanobacteria, all other photosynthetic organisms are prokaryotes. Thus, this key fundamental innovation which either directly or indirectly sustains all eukaryotic organisms, originated within the prokaryotes.

Within prokaryotes, photosynthetic capability is present within five major groups of bacteria:

1. *Firmicutes* or the low G+C Gram-positive bacteria (*Heliobacterium*),
2. *Chloroflexi* or Green non-sulphur bacteria,
3. *Chlorobi* or Greensulphur bacteria,
4. Proteobacteria and,
5. Cyanobacteria.

Of these only Cyanobacteria, which contains two different

reaction centres (RC) RC-1 and RC-2 (or PS I and PS II) linked to each other, are capable of carrying out oxygenic photosynthesis. All other photosynthetic bacteria carry out only anoxygenic photosynthesis and contain a single reaction centre. Of these *Heliobacteria* and *Chlorobi* contain Fe-S type of reaction centres (RC-1) whereas *Chloroflexi* and Proteobacteria have a pheophytin-quinone type of reaction centre (RC-2). The similarities of these RCs in component parts and the mechanisms of charge transfer indicate that they have evolved from a common ancestor. To understand the origin of photosynthesis and which of these reaction centres first evolved, it is essential to understand the branching order of different photosynthetic phyla from a common ancestor, which is not resolved by traditional phylogenetic means. However, based upon the signature sequence approach the branching orders of different bacteria phyla can now be reliably deducted.

Earliest Branching Photosynthetic Bacteria

Firmicutes (*Heliobacterium*) are indicated to be earliest branching photosynthetic bacteria. The ancestral nature of this group is also supported by a number of other observations:

1. Unlike other photosynthetic bacteria, both antenna and reaction centre activities are present within a single protein in Heliobacteria;
2. The reaction centre complex in *Heliobacteria* (and also green sulphur bacteria) has a simpler homodimeric structure as opposed to being heterodimeric in other photosynthetic bacteria;
3. The RC in *Heliobacteria* contains a unique photosynthetic pigment Bchl *g*, which is indicated to be primitive in comparison to the pigments found in other photosynthetic organisms.
4. Of the different photosynthetic bacteria, only *Heliobacteria* are bounded by a single unit lipid membrane (monoderm cell structure), which is

indicated to be an ancestral characteristic in comparison to the cells containing both an inner and outer cell membranes (Diderm cell structure).

Second Photosynthetic Bacteria

Following *Heliobacteria, Chloroflexi* are indicated to be the next group of photosynthetic organisms that branched off from the common ancestor.

The branching of both *Heliobacteria* and *Chloroflexi* prior to Cyanobacteria provides evidence that both RC-1 and RC-2 had already evolved prior to the emergence of Cyanobacteria, which contain both of these reactions centres linked to each other.

Anoxygenic Photosynthesis vs Oxygenic Photosynthesis

The bacterial groups utilising anoxygenic photosynthesis mode evolved much earlier than those capable of oxygenic photosynthesis.

This is in accordance with the observation that change in atmosphere from anoxygenic to oxygenic occurred much later (between 1.5-2 billion year) after the evolution of earlier organisms. This observation indicates that the earlier prokaryotic fossils probably do not correspond to Cyanobacteria but some other groups of photosynthetic bacteria.

Later Branching Photosynthetic Bacteria

The later branching photosynthetic phyla which contain either one or both of these RCs could have acquired such genes from the earlier branching lineages by either direct descent or by means of lateral gene transfer.

Speculations about the Earliest Organism

The presence of photosynthetic ability in the earliest branching bacterial phylum indicates that photosynthesis evolved very early in evolution and it is possible that the earliest organism.

Anoxygenic Photosynthetic Organisms

Some photosynthetic bacteria can use light energy to extract electrons from molecules other than water. These organisms are of ancient origin, presumed to have evolved before oxygenic photosynthetic organisms. Anoxygenic photosynthetic organisms occur in the domain bacteria and have representatives in four phyla Purple Bacteria, Green Sulphur Bacteria, Green Gliding Bacteria, and Gram Positive Bacteria. The phylum of the proteobacteria, defined on the basis of 16S r-RNA analysis, forms a large and diverse group of phototrophic and non-phototrophic, chemoheterotrophic, gram-negative organisms.

The proteobacteria are subdivided into five subgroups, labelled alpha to epsilon. Alpha, beta, and gamma—subgroups contain photosynthetic as well as non-photosynthetic representatives; for the delta and epsilon subgroups, photosynthetic species are so far unknown. The delta and epsilon subgroups use menaquinone as their pool quinone (just as the large majority of the bacteria) whereas alpha, beta and gamma—proteobacteria have electron transfer chains based on ubiquinone.

The transition from the low-potential (MK-) electron transfer chains towards the high-potential (UQ-) chains thus apparently occurred between the delta and the gamma-subgroups. A comparable rise in redox poise of the electron transfer chains occurred on the cyanobacterial branch. Purple bacteria are distinguished into the Chromataceae (formerly named Thiorhodaceae) and the Rhodospirillaceae (formerly called Athiorhodaceae). Purple bacteria perform an anoxygenic type II- photosynthesis which was until recently considered to be switched on exclusively under anaerobic conditions. However, several representatives of aerobic phototrophic purple bacteria have recently been isolated.

These aerobic phototrophs are frequently considered to represent an evolutionary transition from phototrophicity towards chemoheterotrophicity within the proteobacteria.

The electron transport chain of purple bacteria minimally consists of an RC2, a cytochrome bc1-complex, the quinone pool and a small electron carrier mediating between the bc1-complex and the RC2. In the majority of cases, the RC2 additionally contains a tetrahaem cytochrome serving as immediate electron donor to the photo-oxidised pigment. Intriguingly, a similar tetrahaem subunit is present in the RC2-type photosystem of the phylogenetically distant green filamentous bacteria (Chloroflexaceae). This has been interpreted as indicating an ancient origin of the family of RC2-type photosystems.

Previously, it had been taken for granted that cytochrome c2 plays the role of the soluble electron carrier between the cytochrome bc1-complex and the RC2 in purple bacteria in general. During recent years, it has been demonstrated that:

(a) Membrane-bound cytochromes can substitute for cytochrome c2 (especially the so-called "*cy*-group") and that

(b) The role of the soluble electron shuttle is played by a High Potential Iron Sulphur Protein (HiPIP) in a large number of purple bacteria.

Photosynthetic Reaction Centre

Most of the processes in the inorganic sulphur metabolism of prokaryotes, especially those involving elemental sulphur, are poorly described despite their significance for cellular growth and their impact on the environment. This incomplete knowledge is in part due to the limitations posed by traditional genetic and enzymological approaches.

We investigate the inorganic sulphur metabolism in the green sulphur bacterium *Chlorobium tepidum* based on a genomics approach as a model for the inorganic sulphur metabolism in both phototrophic and chemotrophic prokaryotes. Based on our genome analyses and the limited biochemical and physiological information available on the inorganic sulphur metabolism in green sulphur bacteria, it is obvious that novel enzymes and pathways remain to be discovered. Our research approaches this problem by combining bioinformatic analyses with targeted gene inactivations, transcriptome analyses, and biochemical and physiological studies. Green sulphur bacteria are photosynthetic organisms found in anaerobic and sulphide-containing freshwater and estuarine environments including lakes, sediments, and microbial mats.

They are essential for the natural cycling of sulphur because they oxidise inorganic sulphur compounds such as sulphide, polysulphides, elemental sulphur, and thiosulphate under anaerobic conditions.

Green sulphur bacteria thus contribute significantly to the biogeochemical cycling of carbon, nitrogen, and sulphur. Bacteriochlorophyll c, the light-harvesting pigment from the green sulphur bacteria Chlorobium limicola, has been resolved into over a dozen chemically similar components by a new high pressure reversed-phase chromatographic procedure based on a stationary phase of polyethylene power.

Detailed spectroscopic characterisation of the resolved components has resulted in the identification of four different chlorins and six different esterifying alcohols. The major esterifying alcohol is trans, trans-farnesol, but smaller amounts of geranylgeraniol, tetrahydrogeranylgeraniol, phytol, cis-9-hexadecen-1-ol and 4-undecyl-2-furanmethanol are also observed.

The high information content intrinsic in the compositional analyses of the complex pigment mixtures found in green sulphur bacteria appears to provide a new probe of the mechanisms of pigment biosynthesis in these organisms.

Water Photolysis

The NADPH is the main reducing agent in chloroplasts, providing a source of energetic electrons to other reactions. Its production leaves chlorophyll with a deficit of electrons (oxidised), which must be obtained from some other reducing agent. The excited electrons lost from chlorophyll in photosystem I are replaced from the electron transport chain by plastocyanin. However, since photosystem II includes the first steps of the *Z-scheme,* an external source of electrons is required to reduce its oxidised *chlorophyll a* molecules. The source of electrons in green-plant and cyanobacterial photosynthesis is water.

Two water molecules are oxidised by four successive charge-separation reactions by photosystem II to yield a molecule of diatomic oxygen and four hydrogen ions; the electron yielded in each step is transferred to a redox-active tyrosine residue that then reduces the photo-oxidised paired-chlorophyll *a* species called P680 that serves as the primary (light-driven) electron donor in the photosystem II reaction centre. The oxidation of water is catalysed in photosystem II by a redox-active structure that contains four manganese ions; this oxygen-evolving complex binds two water molecules and stores the four oxidising equivalents that are required to drive the water-oxidising reaction. Photosystem II is the only known biological enzyme that carries out this oxidation of water. The hydrogen ions contribute to the transmembrane chemiosmotic potential that leads to ATP synthesis. Oxygen is a waste product of light-independent reactions, but the majority of organisms on Earth use oxygen for cellular respiration, including photosynthetic organisms.

Photodissociation

Photodissociation (or photolysis) is a chemical reaction in which a chemical compound is broken down by photons. Photodissociation is not limited to visible light, but to have enough energy to break up a molecule; the photon is likely to be an electromagnetic wave with the energy of visible light or higher, such as ultraviolet light, x-rays and gamma rays. The direct process is defined as the interaction of one photon interacting with one target molecule.

Photolysis in Photosynthesis

Photolysis is a part of the light-dependent reactions of photosynthesis. The general reaction of photosynthetic photolysis can be given as:

$$H_2A + 2 \text{ photons (light)} \rightarrow 2e^- + 2H^+ + A$$

The chemical nature of "A" depends on the type of organism. For example in purple sulphur bacteria, hydrogen sulphide (H_2S) is oxidised to sulphur (S). In oxygenic photosynthesis, water (H_2O) serves as a substrate for photolysis resulting in the generation of free oxygen (O_2). This process is responsible for generating the majority of breathable oxygen in earth's atmosphere. Photolysis of water occurs in the thylakoids of cyanobacteria and the chloroplasts of green algae and plants.

The effectiveness of photons of different wavelengths depends on the absorption spectra of the photosynthetic pigments in the organism. Chlorophylls absorb light in the violet-blue and red parts of the spectrum, while accessory pigments capture other wavelengths as well. The phycobilins of red algae absorb blue-green light which penetrates deeper into water than red light, enabling them to photosynthesise in deep waters. Each absorbed photon causes the formation of an exciton (an electron excited to a higher energy state) in the pigment molecule. The energy of the exciton is transferred to a chlorophyll a molecule (P680, where P stands for pigment and 680 for its absorption maximum at 680 nm) in the reaction centre of photosystem II via resonance energy transfer. P680 can also directly absorb a photon at a suitable wavelength.

Photolysis during photosynthesis occurs in a series of light-driven oxidation events. The energised electron (exciton) of P680 is captured by a primary electron acceptor of the photosynthetic electron transfer chain and thus exits photosystem II. In order to repeat the reaction, the electron in the reaction centre needs to be replenished. This occurs by oxidation of water in the case of oxygenic photosynthesis. The electron-deficient reaction centre of photosystem II (P680*) is the strongest biological oxidising agent known on earth, which allows it to break apart molecules as stable as water.

The water-splitting reaction is catalysed by the oxygen evolving complex of photosystem II. This protein-bound inorganic complex contains four manganese ions, plus a calcium and chloride ion as co-factors. Two water molecules are complexed by the manganese cluster, which then undergoes a series of four electron removals (oxidations) to replenish the reaction centre of photosystem II. At the end of this cycle, free oxygen (O_2) is generated and the hydrogen of the water molecules has been converted to four protons released into the thylakoid lumen.

These protons, as well as additional protons pumped across the thylakoid membrane coupled with the electron transfer chain, form a proton gradient across the membrane that drives photophosphorylation and thus the generation of chemical energy in the form of adenosine triphosphate (ATP). The electrons reach the P700 reaction centre of photosystem I where they are energised again by light.

They are passed down another electron transfer chain and finally combine with the co-enzyme $NADP^+$ and protons outside the thylakoids to NADPH. Thus, the net oxidation reaction of water photolysis can be written as:

$$2H_2O + 2NADP^+ + 8 \text{ photons (light) } 2NADPH + 2H^+ + O_2$$

The free energy change for this reaction is 102 kilocalories per mole. Since the energy of light at 700 nm is about 40 kilocalories per mole of photons, approximately 320 kilocalories of light energy are available for the reaction. Therefore, approximately one-third of the available light energy is captured as NADPH during photolysis and electron transfer. An equal amount of ATP is generated by the resulting proton gradient. Oxygen as a by-product is of no further use to the reaction and thus released into the atmosphere.

Photolysis in Atmosphere

Photolysis also occurs in the atmosphere as part of a series of reactions by which primary pollutants such as hydrocarbons and nitrogen oxides react to form secondary pollutants such as peroxyacyl nitrates.

The two most important photodissociation reactions in the troposphere are firstly:

$$O_3 + h\nu \rightarrow O_2 + O(^1D)\ \lambda < 320\ \text{nm}$$

which generates an excited oxygen atom which can go on to react with water to give the hydroxyl radical:

$$O(^1D) + H_2O \rightarrow 2OH$$

The hydroxyl radical is central to atmospheric chemistry as it initiates the oxidation of hydrocarbons in the atmosphere and so acts like a detergent.

Secondly the reaction:

$$NO_2 + h\nu \rightarrow NO + O$$

is a key reaction in the formation of tropospheric ozone.

The formation of the ozone layer is also caused by photodissociation. Ozone in the earth's stratosphere is created by ultraviolet light striking oxygen molecules containing two oxygen atoms (O_2), splitting them into individual oxygen atoms (atomic oxygen); the atomic oxygen then combines with unbroken O_2 to create ozone, O_3. In addition, photolysis is the process by which CFCs are broken down in the upper atmosphere to form ozone-destroying chlorine free radicals.

Astrophysics

In astrophysics, photodissociation is one of the major processes through which molecules are broken down (but new molecules are being formed). Because of the vacuum of the interstellar medium, molecules and free radicals can exist for a long time. Photodissociation is the main path by which molecules are broken down. Photodissociation rates are very important in the study of the composition of interstellar clouds in which stars are formed.

Typical examples of photodissociation in the interstellar medium are ($h\nu$ is the scientific notation for light, specifically a photon).

Multiple Photon Dissociation

In comparison to ultraviolet or other high energy photons, single photons in the infra-red spectral range usually are not energetic enough for direct photodissociation of molecules. However, after absorption of multiple infra-red photons a molecule may gain internal energy to overcome its barrier for dissociation. Multiple photon dissociation (MPD) can be achieved by applying high power lasers, e.g. a carbon dioxide laser, or a free electron laser, or by long interaction times of the molecule with the radiation field without the possibility for rapid cooling, e.g. by collisions. The latter method allows even for MPD induced by black body radiation.

Oxygen Evolution

Oxygen evolution is the process of generating molecular oxygen through chemical reaction. Mechanisms of oxygen evolution include the photolysis of water during oxygenic photosynthesis, electrolysis of water into oxygen and hydrogen, and electrocatalytic oxygen evolution from oxides and oxoacids.

Oxygen Evolution in Nature

Photolytic oxygen evolution is the fundamental process by which breathable oxygen is generated in earth's biosphere. The reaction is part of the light-dependent reactions of photosynthesis in cyanobacteria and the chloroplasts of green algae and plants. It utilises the energy of light to split a water molecule into its protons and electrons for photosynthesis. Free oxygen is generated as a waste product of this reaction, and is released into the atmosphere.

Biochemical Reaction

Photolytic oxygen evolution occurs via the light-dependent oxidation of water to molecular oxygen and can be written as the following simplified chemical reaction:

$$2H_2O \rightarrow 4e^- + 4H^+ + O_2$$

The reaction requires the energy of four photons. The electrons from the oxidised water molecules replace electrons in the P_{680} component of photosystem II which have been removed into an electron transport chain via light-dependent excitation and resonance energy transfer onto plastoquinone. Photosytem II therefore has also been referred to as water-plastoquinone oxido-reductase. The protons are released into the thylakoid lumen, thus contributing to the generation of a proton gradient across the thylakoid membrane. This proton gradient is the driving force for ATP synthesis via photophosphorylation and coupling the absorption of light energy and photolysis of water to the creation of chemical energy during photosynthesis.

Oxygen-evolving Complex

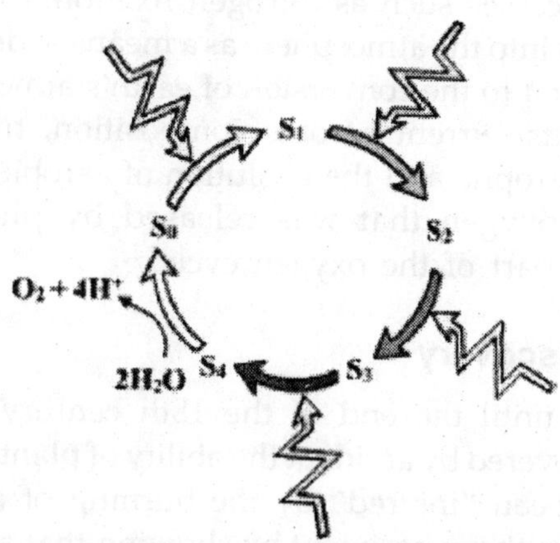

Oxygen evolution by water oxidation during photosynthesis

The jagged lines represent four photons oxidising the central cluster of the oxygen evolving complex by exciting and removing four electrons through a cycle of *S-states*.

Water oxidation is catalysed by a manganese-containing enzyme complex associated with thylakoid membranes known as the oxygen evolving complex (OEC) or water-splitting complex. Manganese is an important co-factor, and calcium and chloride are also required for the reaction to occur.

X-ray crystallography studies have recently provided detailed models of the structure of the oxygen-evolving complex and its manganese cluster. Based on structural and spectroscopic experiments, oxygen evolution involves a core three-plus-one cluster of three manganese ions and one calcium ion, with one additional manganese, which are oxidised via intermediate states called *S-states*. The O-O bond of molecular oxygen is formed between manganese-ligated oxygen atoms at the most oxidised, or S4, state.

Evolution of Oxygen Evolution

Oxygen production during photosynthesis evolved on earth around 2.7 to 3.5 billion years ago. Oxygen was not only a waste product of this reaction, but was also toxic to many metabolic processes such as nitrogen fixation. Consequently, it was released into the atmosphere as a means of detoxification. This contributed to the conversion of earth's atmosphere from anaerobic to its current aerobic composition, triggering the Oxygen Catastrophe and the evolution of aerobic metabolism utilising the oxygen that was released by photosynthetic organisms as part of the oxygen cycle.

History of Discovery

It wasn't until the end of the 18th century that Joseph Priestley discovered by accident the ability of plants to "restore" air that had been "injured" by the burning of a candle. He followed up on the experiment by showing that air "restored" by vegetation was *"not at all inconvenient to a mouse."* He was later awarded a medal for his discoveries that: *"...no vegetable grows in vain... but cleanses and purifies our atmosphere."* Priestley's experiments were followed up by Jan Ingenhousz, a Dutch physician, who showed that "restoration" of air only worked in the presence of light and green plant parts.

Ingenhousz suggested in 1796 that CO_2 (carbon dioxide) is split during photosynthesis to release oxygen, while the carbon combined with water to form carbohydrates. While this hypothesis was attractive and reasonable and thus widely accepted for a long time, it was later proven incorrect. Graduate student C.B. Van Niel at Stanford University found that purple sulphur bacteria reduce carbon to carbohydrates, but accumulate sulphur instead of releasing oxygen. He boldly proposed that in analogy to the sulphur bacteria forming elemental sulphur from H_2S (hydrogen sulphide), plants would form oxygen from H_2O (water). In 1937, this hypothesis was corroborated by the discovery that plants are capable of producing oxygen in the absence of CO_2. This discovery was made by Robin Hill, and subsequently the light-driven release of oxygen in the absence of CO_2 was called the *Hill reaction*. Our current knowledge of the mechanism of oxygen evolution during photosynthesis was further established in experiments tracing isotopes of oxygen from water to oxygen gas.

Technological Oxygen Evolution

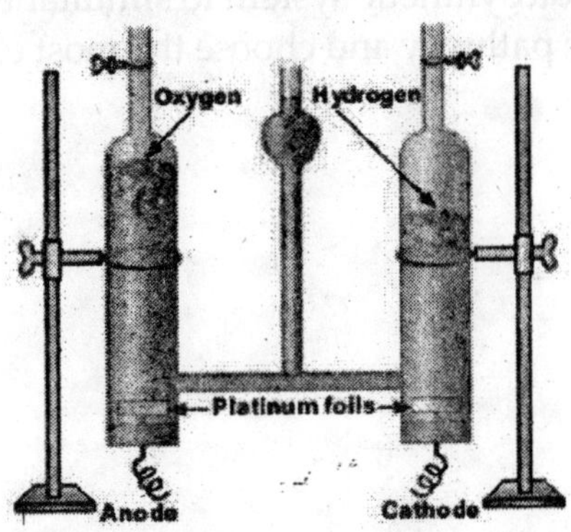

Hoffman electrolysis apparatus used in electrolysis of water

Oxygen evolution occurs as a by-product of hydrogen production via electrolysis of water. While oxygen production is not the main focus of industrial applications of water electrolysis, it becomes essential for life support systems in situations that require the generation of oxygen for air revitalisation. Human exploration of regions that lack breathable oxygen, such as the deep sea or outer space, requires means of reliably generating oxygen apart from earth's atmosphere. Submarines and spacecraft utilise either an electrolytic mechanism (water or solid oxide electrolysis) or chemical oxygen generators as part of their life support equipment.

Quantum Mechanical Effects

Through photosynthesis, sunlight energy is transferred to molecular reaction centres for conversion into chemical energy with nearly 100-per cent efficiency. The transfer of the solar energy takes place almost instantaneously, so little energy is wasted as heat.

A study led by researchers with the US Department of Energy's Lawrence Berkeley National Laboratory (Berkeley Lab) and the University of California at Berkeley suggests that long-lived wavelike electronic quantum coherence plays an important part in this instantaneous transfer of energy by allowing the photosynthetic system to simultaneously try each potential energy pathway and choose the most efficient option.

Photosynthetic Process

Photosynthesis

Photosynthesis is the process of converting light energy to chemical energy and storing it in the bonds of sugar. This process occurs in plants and some algae (Kingdom Protista). Plants need only light energy, CO_2, and H_2O to make sugar. The process of photosynthesis takes place in the chloroplasts, specifically using chlorophyll, the green pigment involved in photosynthesis. Photosynthesis takes place primarily in plant leaves, and little to none occurs in stems, etc. The parts of a typical leaf include the upper and lower epidermis, the mesophyll, the vascular bundle(s) (veins), and the stomates. The upper and lower epidermal cells do not have chloroplasts, thus photosynthesis does not occur there. They serve primarily as protection for the rest of the leaf. The stomates are holes which occur primarily in the lower epidermis and are for air exchange: they let CO_2 in and O_2 out. The vascular bundles or veins in a leaf are part of the plant's transportation system, moving water and nutrients around the plant as needed. The mesophyll cells have chloroplasts and this is where photosynthesis occurs.

As you hopefully recall, the parts of a chloroplast include the outer and inner membranes, intermembrane space, stroma, and thylakoids stacked in grana. The chlorophyll is built into the membranes of the thylakoids.

Chlorophyll looks green because it absorbs red and blue light, making these colours unavailable to be seen by our eyes. It is the green light which is not absorbed that finally reaches our eyes, making chlorophyll appear green. However, it is the energy from the red and blue light that are absorbed that is, thereby, able to be used to do photosynthesis. The green light we can see is not/cannot be absorbed by the plant, and thus cannot be used to do photosynthesis.

The overall chemical reaction involved in photosynthesis is: $6CO_2 + 12H_2O$ (+ light energy) $\rightarrow C_6H_{12}O_6 + 6O_2 + 6H_2O$. This is the source of the O_2 we breathe, and thus, a significant factor in the concerns about deforestation.

There are two parts to photosynthesis: The light reaction happens in the thylakoid membrane and converts light energy to chemical energy. This chemical reaction must, therefore, take place in the light. Chlorophyll and several other pigments such as beta-carotene are organised in clusters in the thylakoid membrane and are involved in the light reaction.

Each of these differently-coloured pigments can absorb a slightly different colour of light and pass its energy to the central chlorophyll molecule to do photosynthesis. The central part of the chemical structure of a chlorophyll molecule is a porphyrin ring, which consists of several fused rings of carbon and nitrogen with a magnesium ion in the centre.

The energy harvested via the light reaction is stored by forming a chemical called ATP (adenosine triphosphate), a compound used by cells for energy storage. This chemical is made of the nucleotide adenine bonded to a ribose sugar, and that is bonded to three phosphate groups. This molecule is very similar to the building blocks for our DNA.

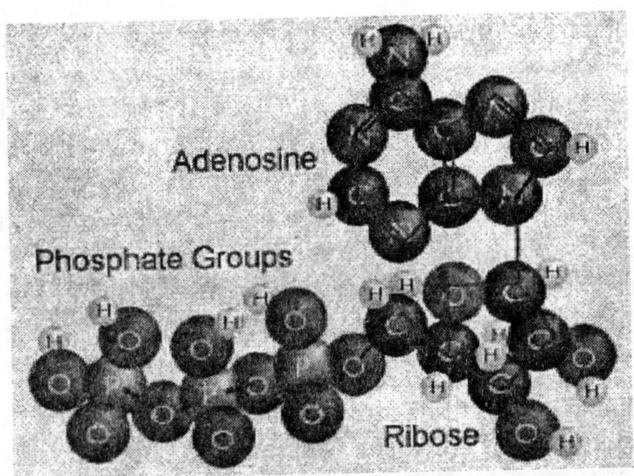

The dark reaction takes place in the stroma within the chloroplast, and converts CO_2 to sugar. This reaction doesn't directly need light in order to occur, but it does need the products of the light reaction (ATP and another chemical called NADPH). The dark reaction involves a cycle called the Calvin cycle in which CO_2 and energy from ATP are used to form sugar. Actually, notice that the first product of photosynthesis is a three-carbon compound called glyceraldehyde 3-phosphate. Almost immediately, two of these join to form a glucose molecule.

Calvin Cycle and Enzymatic Process

Most plants put CO_2 directly into the Calvin cycle. Thus the first stable organic compound formed is the glyceraldehyde 3-phosphate. Since that molecule contains three carbon atoms, these plants are called C_3 plants. For all plants, hot summer weather increases the amount of water that evaporates from the plant. Plants lessen the amount of water that evaporates by keeping their stomates closed during hot, dry weather. Unfortunately, this means that once the CO_2 in their leaves reaches a low level, they must stop doing photosynthesis. Even if there is a tiny bit of CO_2 left, the enzymes used to grab it and put it into the calvin cycle just don't have enough CO_2 to use. Typically the grass in our yards just turns brown and goes dormant. Some plants like crabgrass, corn, and sugar cane have a special modification to conserve water. These plants capture CO_2 in a different way: they do an extra step first, before doing the Calvin cycle. These plants have a special enzyme that can work better, even at very low CO_2 levels, to grab CO_2 and turn it first into oxaloacetate, which contains four carbons. Thus, these plants are called C_4 plants. The CO_2 is then released from the oxaloacetate and put into the Calvin cycle. This is why crabgrass can stay green and keep growing when all the rest of your grass is dried up and brown.

There is yet another strategy to cope with very hot, dry, desert weather and conserve water. Some plants (for example, cacti and pineapple) that live in extremely hot, dry areas like deserts, can only safely open their stomates at night when the weather is cool. Thus, there is no chance for them to get the CO_2 needed for the dark reaction during the daytime. At night when they can open their stomates and take in CO_2, these plants incorporate the CO_2 into various organic compounds to store it. In the daytime, when the light reaction is occurring and ATP is available (but the stomates must remain closed), they take the CO_2 from these organic compounds and put it into the Calvin cycle. These plants are called CAM plants, which stands for crassulacean acid metabolism after the plant family, crassulaceae (which includes the garden plant *Sedum*) where this process was first discovered.

The energy conversion/storage process called photosynthesis occurs in autotrophic plants within organelle called chloroplast. This organelle is double membrane bound structure; necessary pigments are concentrated within inner membrane of chloroplast.

Inner membrane represented as a series of staked membranes = *Granum* (*Grana* = plural). Each Granum is made up of a number of individual membrane units called a *Thylakoid*. The portion of photosynthesis which is called the *Light Reaction* (*Light Dependent* Rx—requires light energy in order to occur) takes place in the Grana.

The clear, non-membrane areas of chloroplast = *Stroma*. This portion of photosynthesis is called the *Dark Reaction* (*Light Independent*—does not require light energy in order to take place) occurs in the Stroma).

Photosynthesis is responsive to certain wave lengths of light energy. The visible spectrum, from red to violet light is effective in photosynthesis The most effective colours of light energy are in the red and blue/violet range of the colour spectrum. These are the wavelengths of light that are absorbed most strongly by plant pigments, particularly the *Primary* (and dominant) pigment in photosynthesis, *Chlorophyll* a.

In addition to chlorophyll a, there are a number of other pigments concentrated within the membranes of the grana. These are called *Accessory Pigments*. All of these *Accessory* pigments are responsible for capturing (absorbing) specific wavelengths (colours) of light energy and transfering this energy to chlorophyll a.

Recall the ultimate goal of photosynthesis (i.e., the end product) is to produce a form of chemical energy (glucose) which can be eventually be converted to large quantities of ATP during cell respiration. The primary function of the Light Rx is to capture and convert light energy to another form of energy useful in a part of the photosysnthesis process, and ultimately to produce glucose.

This portion of photosynthesis (*Light Rx*) is sometimes referred to as the light dependent Rx, in that light energy is always required for it to proceed. It is also know as the Hill Rx. This part of photosynthesis is divided into two phases, or *Systems*, called *Photosystem II* and *Photosystem*.

Structure of Chlorophyll

Chlorophyll a, is concentrated within the membranes of the grana associated with Photosystem II. As red and blue/violet light energy is absorbed by chlorophyll a, the electrons in the outer shell of this pigment molecule are excited and raised to a higher energy level. These electrons are captured by an electron acceptor molecule, and transferred through a series of electron acceptor molecules (as part of an electron transport system) within the grana.

At one point along the pathway of this electron transfer process, the electron transfer provides a sufficient drop in energy potential ("release of energy") to allow for the production of ATP from ADP and P (i). Thus, light energy has been transferred to ATP, which will carry this energy to the *Dark Rx* for production of glucose.

As the electrons in chlorophyll a are excited and passed between the two photosystems producing ATP, these "electron holes, or "lost" electrons from chlorophyll a within photosystem II must be replaced. In that there is no difference between electrons and hydrogen, water will provide a ready source of hydrogen to full these "electron holes". Thus, water molecules are split ("cleaved") and a portion of the hydrogen are used to replace these "lost" electrons. The remainder of the water molecule is recombined to form oxygen—one of the products of photosynthesis, and a specific product of the light Rx.

Pigment associated with photosystem I is a unique form of chlorophyll a called P 700. This pigment is excited by wavelengths (colours) of light energy in the *Far Red* (just longer than the visible red light) portion of the colour spectrum. Electrons from P 700 are picked up by another electron acceptor molecule, and ultimately produces NADPH from NADP+. NADPH is an electron (hydrogen) carrier molecule, which is also used in the DARK Rx in the process of producing glucose. NADPH essentially provides the hydrogen which are part of the glucose molecule. The electrons which are "lost" from P 700 are replaced by the electrons flowing from the electron transport chain linking Photosystem II with Photosystem I.

The portion of photosynthesis takes place in the *Stroma* (clear areas of the chloroplast) and does not require light energy to proceed, but does require the products the (ATP and

NADPH) of the light Rx. These are sometimes referred to as the "driving force" behind the Dark Rx. The Dark Rx is also referred to as the Calvin Cycle or the Carbon Fixation Cycle. This is where carbon dioxide is used in the photosynthesis process, and the CO_2 provides the carbon atom which is the structural backbone of the glucose molecule.

In the molecular structure of glucose, C6H12O6, note that the carbon is provided by the CO_2 in the Dark Rx. The hydrogen is provided by the NADPH, which is produced in the Light Rx and carried to the Dark Rx; the stored chemical energy as part of the glucose molecule is provided by the ATP (from the Light Rx).

The primary source of energy for nearly all life is the Sun. The energy in sunlight is introduced into the biosphere by a process known as photosynthesis, which occurs in plants, algae and some types of bacteria. Photosynthesis can be defined as the physico-chemical process by which photosynthetic organisms use light energy to drive the synthesis of organic compounds. The photosynthetic process depends on a set of complex protein molecules that are located in and around a highly organised membrane. Through a series of energy transducing reactions, the photosynthetic machinery transforms light energy into a stable form that can last for hundreds of millions of years.

Photosynthesis is the physico-chemical process by which plants, algae and photosynthetic bacteria use light energy to drive the synthesis of organic compounds. In plants, algae and certain types of bacteria, the photosynthetic process results in the release of molecular oxygen and the removal of carbon dioxide from the atmosphere that is used to synthesise carbohydrates (oxygenic photosynthesis). Other types of bacteria use light energy to create organic compounds but do not produce oxygen (anoxygenic photosynthesis). Photosynthesis provides the energy and reduced carbon required for the survival of virtually all life on our planet, as well as the molecular oxygen necessary for the survival of oxygen consuming organisms. In addition, the fossil fuels currently being burned to provide energy for human activity were produced by ancient photosynthetic organisms. Although photosynthesis occurs in cells or organelles that are typically only a few microns across, the process has a profound impact on the earth's atmosphere and climate. Each year more than 10 per cent of the total atmospheric carbon dioxide is reduced to carbohydrate by photosynthetic organisms. Most, if not all, of the reduced carbon is returned to the atmosphere as carbon dioxide by microbial, plant and animal metabolism, and by biomass combustion. In turn, the performance of photosynthetic organisms depends on the earth's atmosphere and climate. The large increase in the amount of atmospheric carbon dioxide created by human activity is certain to have a profound impact on the performance and competition of photosynthetic organisms. Knowledge of the physico-chemical process of photosynthesis is essential for understanding the relationship between living organisms and the atmosphere and the balance of life on earth.

Specific Features of Photosynthesis

The overall equation for photosynthesis is deceptively simple. In fact, a complex set of physical and chemical reactions must occur in a coordinated manner for the synthesis of carbohydrates. To produce a sugar molecule such as sucrose, plants require nearly 30 distinct proteins that work within a complicated membrane structure. Research into the mechanism of photosynthesis centres on understanding the structure of the photosynthetic components and the molecular processes that use radiant energy to drive carbohydrate synthesis. The research involves several disciplines, including physics, biophysics, chemistry, structural biology, biochemistry, molecular biology and physiology, and serves as an outstanding example of the success of multidisciplinary research. As such, photosynthesis presents a special challenge in understanding several interrelated molecular processes.

Leaf Structure and Photosynthesis

The Structure of a Leaf and Chloroplast: The chloroplast is bounded by a double membrane: the outer membrane contains proteins that render it permeable to small molecules (MW < 6000); the inner membrane forms the permeability barrier of the organelle. Photosynthesis occurs on the thylakoid membrane, which forms a series of flattened vesicles (thylakoids) enclosing a single interconnected luminal space. The green colour of plants is due to the green colour of chlorophyll, all of which is localised to the thylakoid membrane. A granum is a stack of adjacent thylakoids. The stroma is the space within the inner membrane surrounding the thylakoids.

Plants are the only photosynthetic organisms to have leaves (and not all plants have leaves). A leaf may be viewed as a solar collector crammed full of photosynthetic cells.

The raw materials of photosynthesis, water and carbon dioxide, enter the cells of the leaf, and the products of photosynthesis, sugar and oxygen, leave the leaf.

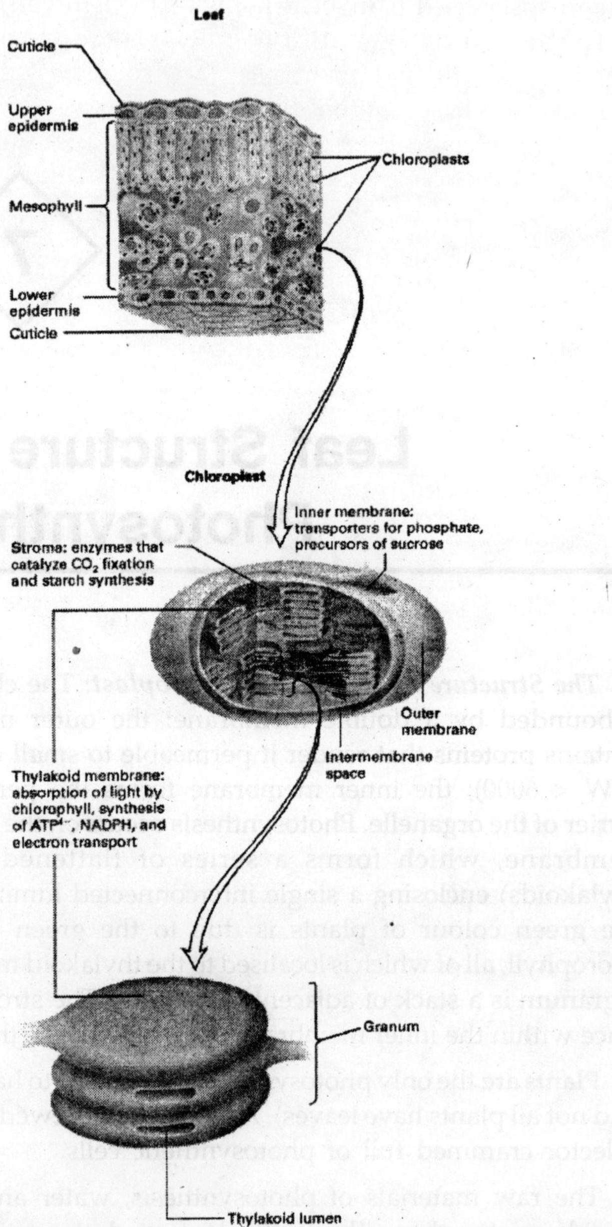

Water enters the root and is transported up to the leaves

through specialised plant cells known as xylem vessels. Land plants must guard against drying out and so have evolved specialised structures known as *stomata* to allow gas to enter and leave the leaf. Carbon dioxide cannot pass through the protective waxy layer covering the leaf (cuticle), but it can enter the leaf through the stoma (the singular of stomata), flanked by two guard cells. Likewise, oxygen produced during photosynthesis can only pass out of the leaf through the opened stomata. Unfortunately for the plant, while these gases are moving between the inside and outside of the leaf, a great deal of water is also lost. Cottonwood trees, for example, will lose 100 gallons (about 450 dm^3) of water per hour during hot desert days.

Cellular Respiration

Cellular respiration is a process that describes the metabolic reactions and processes that take place in a cell to obtain chemical energy from fuel molecules. Energy is released by the oxidation of fuel molecules and is stored as "high-energy" carriers. The reactions involved in respiration are catabolic reactions in metabolism.

Fuel molecules commonly used by cells in respiration include glucose, amino acids and fatty acids, and a common oxidising agent (electron acceptor) is molecular oxygen (O_2). There are organisms, however, that can respire using other organic molecules as electron acceptors instead of oxygen. Organisms that use oxygen as a final electron acceptor in respiration are described as aerobic, while those that do not are referred to as anaerobic.

The energy released in respiration is used to synthesise molecules that act as a chemical storage of this energy. One of the most widely used compounds in a cell is adenosine triphosphate (ATP) and its stored chemical energy can be used for many processes requiring energy, including biosynthesis, locomotion or transportation of molecules across cell membranes. Because of its ubiquitous nature, ATP is also known as the "universal energy currency", since the amount of it in a cell indicates how much energy is available for energy-consuming processes.

Aerobic Respiration

Aerobic respiration requires oxygen in order to generate energy (ATP). It is the preferred method of pyruvate breakdown from glycolysis and requires that pyruvate enter the mitochondrion to be fully oxidised by the Krebs cycle. The product of this process is energy in the form of ATP (Adenosine Triphosphate), by substrate-level phosphorylation, NADH and FADH2. The reducing potential of NADH and FADH2 is converted to more ATP via an electron transport chain with oxygen as the "terminal electron acceptor". Most of the ATP produced by cellular respiration is by oxidative phosphorylation, ATP molecules are made due to the chemiosmotic potential driving ATP synthase. Respiration is the process by which cells obtain energy when oxygen is present in the cell.

Theoretically, 36 ATP molecules can be made per glucose during cellular respiration, however, such conditions are generally not realised due to such losses as the cost of moving pyruvate into mitochondria. Aerobic metabolism is more efficient than anaerobic metabolism (which yields 2 mol ATP per 1 mol glucose). They share the initial pathway of glycolysis but aerobic metabolism continues with the Krebs cycle and oxidative phosphorylation. The post-glycolytic reactions take place in the mitochondria in eukaryotic cells, and in the cytoplasm in prokaryotic cells.

Localisation of Pyruvate Decarboxylation

In eukaryotic cells the pyruvate decarboxylation occurs inside the mitochondria, after transport of the substrate, pyruvate, from the cytosol. The transport of pyruvate into the mitochondria is via a transport protein and is active, consuming energy. Passive diffusion of pyruvate into the mitochondria is impossible because it is a polar molecule.

On entry to the mitochondria the pyruvate decarboxylation occurs, producing acetyl CoA. This irreversible reaction traps the acetyl CoA within the mitochondria (there is no transporter for acetyl CoA). The carbon dioxide produced by this reaction is non-polar and small, and can diffuse out of the mitochondria and out of the cell. In prokaryotes, which have no mitochondria, this reaction is either carried out in the cytosol, or not at all.

Post-pyruvate Decarboxylation Processes

The acetyl CoA produced by this reaction may go on to a variety of different metabolic pathways. The major usage is the citric acid cycle and aerobic respiration, but acetyl CoA is also a major substrate for lipid and amino acid synthesis. Indirectly, intermediates in the citric acid pathway may also be used for synthesis.

The NADH produced may also be used in several ways. Under aerobic conditions, NADH may be oxidised by the electron transport chain into NAD^+, renewing this reactant for use in oxidative decarboxylation (this requires oxygen). In anaerobic conditions, NAD^+ can be regenerated by anaerobic respiration; however, acetyl CoA will quickly build up as it is no longer consumed by the stalled citric acid cycle, and this inhibits the forward reaction.

Gluconeogenesis

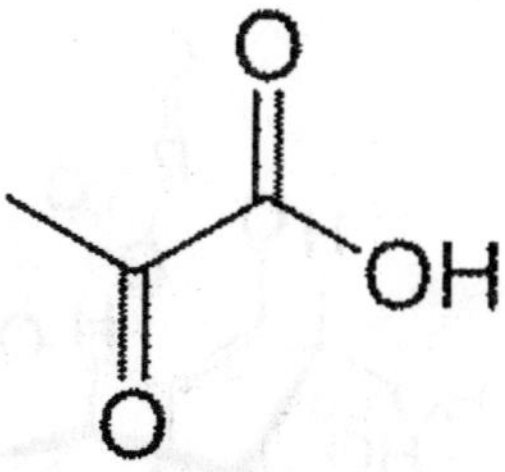

O

OH

O

Pyruvic acid

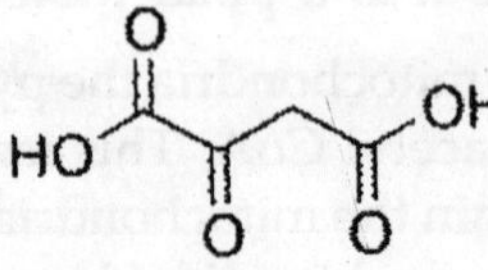

Oxaloacetic acid

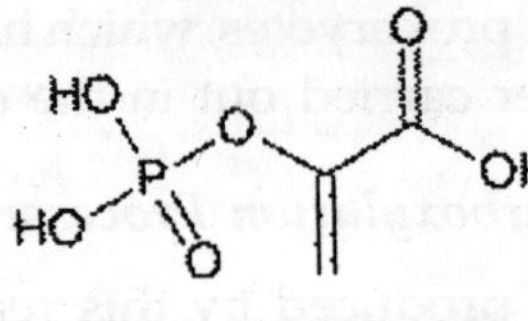

Phosphoenolpyruvate

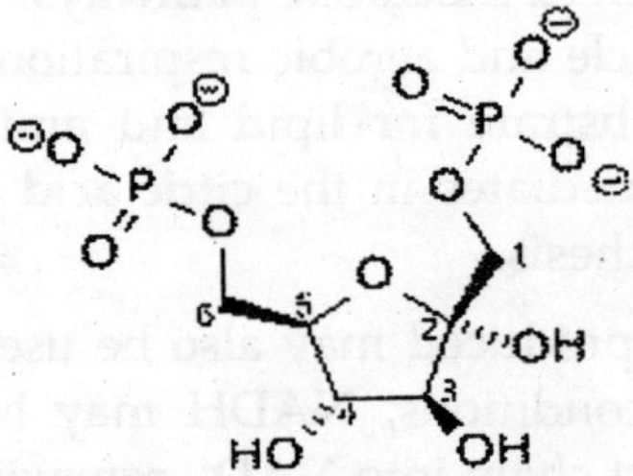

Fructose 1,6-bisphosphate

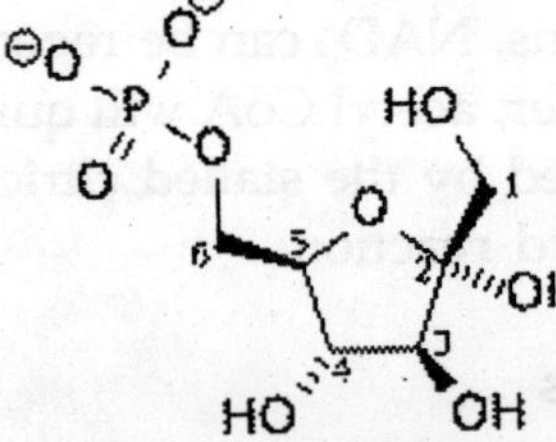

Fructose 6-phosphate

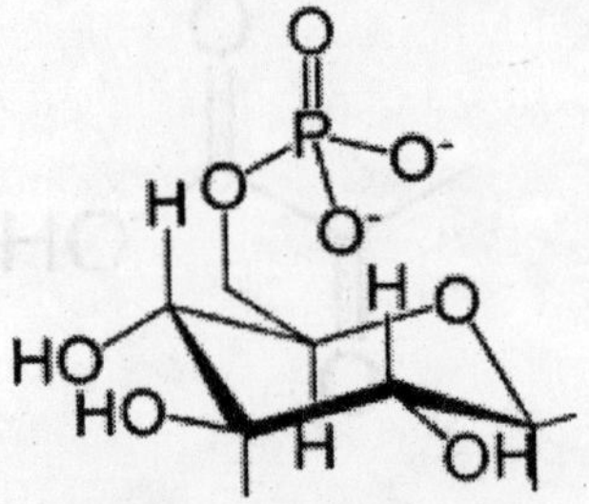

Glucose-6-phosphate

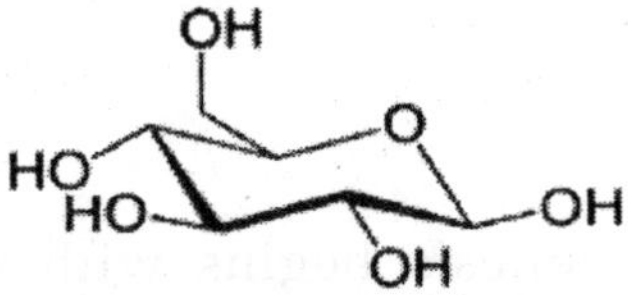

Glucose

Gluconeogenesis is the generation of glucose from non-sugar carbon substrates like pyruvate, lactate, glycerol, and glucogenic amino acids (primarily alanine and glutamine).

The vast majority of gluconeogenesis takes place in the liver and, to a smaller extent, in the cortex of kidney. This process occurs during periods of fasting, starvation, or intense exercise and is highly endergonic.

Entering the Pathway

Many 3- and 4-carbon substrates can enter the gluconeogenesis pathway. Lactate from anaerobic respiration in skeletal muscle is easily converted to pyruvate in the liver cells; this happens as part of the Cori cycle. However, the first designated substrate in the gluconeogenic pathway is pyruvate. Oxaloacetate (an intermediate in the citric acid cycle) can also be used for gluconeogenesis. The gluconeogenic pathway can also generate glucose from amino acids, such as alanine, aspartate, glutamate, or others. Following removal of the amino group (by transamination or deamination) from the amino acid, the remaining carbon skeleton can enter gluconeogenesis directly (as pyruvate or oxaloacetate), or indirectly, e.g., via the citric acid cycle, converting α -ketoglutarate to oxaloacetate.

Most fatty acids cannot be converted into glucose unless the glyoxylate cycle is used, the exception being odd-chain fatty acids, which can yield propionyl CoA, a precursor for succinyl CoA. Fatty acids are regularly broken down into the two-carbon acetyl CoA, which becomes degraded in the citric acid cycle. In contrast, glycerol, which is a part of all triacylglycerols, can be used in gluconeogenesis. In organisms in which glycerol is derived from glucose (e.g., humans and other mammals), glycerol is sometimes not considered a true gluconeogenic substrate, as it cannot be used to generate *new* glucose.

Pathway

- Gluconeogenesis begins with the formation of oxaloacetate through carboxylation of pyruvate at the expense of one molecule of ATP, but is inhibited in the presence of high levels of ADP. This reaction is catalysed by pyruvate carboxylase.
- Oxaloacetate is then decarboxylated and simultaneously phosphorylated by phosphoenolpyruvate carboxy-kinase to produce phosphoenolpyruvate. One molecule of GTP is hydrolysed to GDP in the course of this reaction. Both reactions take place in mitochondria. Oxaloacetate has to be transformed into malate in order to be transported out of the mitochondria.
- The next steps in the reaction are the same as reversed glycolysis. However fructose-1,6-bisphosphatase converts fructose-1,6-bisphosphate to fructose-6-phosphate. The purpose of this reaction is to overcome the large negative Δ G.
- Glucose-6-phosphate is formed from fructose-6-phosphate by phosphoglucoisomerase. Glucose-6-phosphate is used in other pathways. Free glucose is not generated automatically because glucose, unlike glucose-6-phosphate, tends to freely diffuse out of the cell.

  The reaction of actual glucose formation is carried out in the lumen of the endoplasmic reticulum. Here, glucose-6-phosphate is hydrolysed by glucose-6-phosphatase, the last enzyme in gluconeogenesis, to produce glucose. Glucose is then shuttled into the cytosol by glucose transporters located in the membrane of the endoplasmic reticulum.

Regulation

Gluconeogenesis cannot be considered to be simply a reverse process of glycolysis, as the three irreversible steps in glycolysis are bypassed in gluconeogenesis. This is done to ensure that glycolysis and gluconeogenesis are not operating at the same time in the cell, making it a futile cycle. Therefore, glycolysis and gluconeogenesis follow *reciprocal regulation,* that is, cellular conditions, which inhibit glycolysis, may in turn activate gluconeogenesis.

Glucose-6-phosphate regulates the enzyme glucose-6-phosphatase in the lumen of ER by inducing its activity. In contrast, its accumulation will feed-back inhibit hexokinase in glycolysis. Once again, it follows the principle of *reciprocal regulation.*

The majority of the enzymes responsible for gluconeogenesis are found in the cytoplasm; the exceptions are mitochondrial pyruvate carboxylase and mitochondrial phosphoenolpyruvate carboxykinase which are located in the mitochondria. The rate of gluconeogenesis is ultimately controlled by the action of a key enzyme, fructose-1,6-bisphosphatase, which is also regulated through signal transduction by cAMP and its phosphorylation.

Most factors that regulate the activity of the gluconeogenesis pathway do so by inhibiting the activity or expression of key enzymes. However, both acetyl CoA and citrate activate gluconeogenesis enzymes (pyruvate carboxylase and fructose-1,6-bisphosphatase, respectively). Notably, acetyl-CoA and citrate also play inhibitory roles in pyruvate kinase activity in glycolysis.

Hexokinase

Hexokinase is inhibited by glucose-6-phosphate (G6P), the product it forms through the ATP driven phosphorylation. This is necessary to prevent an accumulation of G6P in the cell when flux through the glycolytic pathway is low. Glucose will enter the cell but since the hexokinase is not active it can readily diffuse back to the blood through the glucose transporter in the plasma membrane. If hexokinase remained active during low glycolytic flux the G6P would accumulate and the extra solute would cause the cells to enlarge due to osmosis.

In animals regulation of blood glucose levels by the liver is a vital part of homoeostasis. In liver cells, any extra G6P is stored as glycogen. In these cells hexokinase is not expressed, instead glucokinase catalyses the phosphorylation of glucose to G6P. This enzyme is not inhibited by high levels of G6P and glucose can still be converted to G6P and then be stored as glycogen. This is important when blood glucose levels are high. During hypoglycemia the glycogen can be converted back to G6P and then converted to glucose by a liver specific enzyme glucose 6-phosphatase. This reverse reaction is an important role of liver cells to maintain blood sugars levels during fasting. This is critical for neuron function since they can only use glucose as an energy source.

Phosphofructokinase

Phosphofructokinase is an important control point in the glycolytic pathway since it is immediately downstream of the entry points for hexose sugars.

High levels of ATP inhibit the PFK enzyme by lowering its affinity for F6P. ATP causes this control by binding to a specific regulatory site that is distinct from the catalytic site. This is a good example of allosteric control. AMP can reverse the inhibitory effect of ATP. A consequence is that PFK is tightly controlled by the ratio of ATP/AMP in the cell. This makes sense since these molecules are direct indicators of the energy charge in the cell.

Since glycolysis is also a source of carbon skeletons for biosynthesis, a negative feedback control to glycolysis from the carbon skeleton pool is useful. Citrate is an example of a metabolite that regulates phosphofructokinase by enhancing the inhibitory effect of ATP. Citrate is an early intermediate in the citric acid cycle, and a high level means that biosynthetic precursors are abundant.

Low pH also inhibits phosphofructokinase activity and prevents the excessive rise of lactic acid during anaerobic conditions that could otherwise cause a drop in blood pH (acidosis), (a potentially life threatening condition).

Fructose 2,6-bisphosphate (F2,6BP) is a potent activator of phosphofructokinase (PFK-1) that is synthesised when F6P is phosphorylated by a second phosphofructokinase (PFK2). This second enzyme is inactive when cAMP is high, and links the regulation of glycolysis to hormone activity in the body.

Both glucagon and adrenalin cause high levels of cAMP in the liver. The result is lower levels of liver fructose 2,6-bisphosphate such that gluconeogenesis (glycolysis in reverse) is favoured. This is consistent with the role of the liver in such situations since the response of the liver to these hormones is to releases glucose to the blood.

Pyruvate Kinase and Phosphoglycerate Kinase

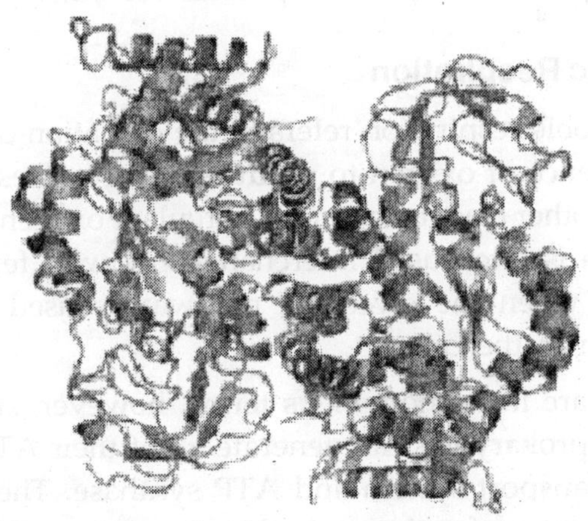

Figure: Yeast pyruvate kinase.

Pyruvate kinase and phosphoglycerate kinase catalyse the two substrate-level phosphorylation steps, and produce ATP from ADP. The requirement of ADP to carry out this reaction provides regulation as when the cell has plenty of ATP it will have little ADP so this reaction is unable to happen. ATP decays relatively quickly, even when not used as an energy source, these stages provide the required simple and fast regulation of ATP levels.

This control is accentuated as, after the formation of F1,6bP, many of the glycolysis reactions are energetically unfavourable. The only reactions that are favourable are these two substrate-level phosphorylation steps. These two reactions pull the glycolytic pathway to completion when ADP is low and ATP is required.

Post-glycolysis Processes

The ultimate fate of pyruvate and NADH produced in glycolysis depends upon the organism and the conditions, most notably the presence or absence of oxygen and other external electron acceptors. In addition, not all carbon entering the pathway leaves as pyruvate and may be extracted at earlier stages to provide carbon compounds for other pathways.

Anaerobic Respiration

Anaerobic respiration refers to the oxidation of molecules in the absence of oxygen to produce energy. These processes require another electron acceptor to replace oxygen. Anaerobic respiration is often used interchangeably with fermentation, especially when the glycolytic pathway is used for energy production in the cell.

]They are not synonymous terms, however, since certain anaerobic prokaryotes can generate all of their ATP using an electron transport system and ATP synthase. The word and symbol equation for the anaerobic respiration of glucose is:

$$\text{Glucose} \longrightarrow \text{Lactic acid} + \text{Energy (ATP)}$$

$$C_6H_{12}O_6 \longrightarrow 2C_3H_6O_3 + 2\ \text{ATP}$$

The energy released is about 120kJ per mole Glucose.

Obligate Anaerobes

In some organisms called *obligate (strict) anaerobes* (ex: *C. tetani* (causes tetanus), *C. perfringens* (causes gangrene)), the presence of oxygen is lethal. This is because the presence of oxygen is processed by the organisms into the extremely toxic molecules of single oxygen ($^1O_2$), superoxide ion (O_2^-), hydrogen peroxide (H_2O_2), hydroxyl ion (OH^-), and other toxic molecules.

Faculative Anaerobes and Obligate Aerobes

Faculative anaerobes (organisms that can survive in either oxygenated or deoxygenated environments and can switch between cellular respiration or fermentation, respectively) and *obligate (strict) aerobes* (organisms that can survive only with oxygen) have special enzymes (superoxide dimutase and catalase) that can safely handle these products and transform them into harmless water and diatomic oxygen in the following reactions:

1. $2O_2^- + 2H^+$ —Superoxide Dismutase⟶ H_2O_2 (hydrogen peroxide) + O_2

The hydrogen peroxide produced is then transferred to a second reaction...

2. $2H_2O_2$ —Catalase⟶ $2H_2O + O_2$

The oxidative powers of the superoxide ion have now been neutralised. Only facultative anaerobes and obligate aerobes possess the two enzymes necessary to reduce the superoxide.

In organisms which use glycolysis, the absence of oxygen prevents pyruvate from being metabolised to CO_2 and water via the citric acid cycle and the electron transport chain (which relies on O_2) does not function. Fermentation does not yield more energy than that already obtained from glycolysis (2 ATPs) but serves to regenerate NAD^+ so glycolysis can continue. Various end products can also be created, such as lactate or ethanol.

Fermentation in animals is essential to human life.

In lactic acid fermentation, the following reaction occurs:

1. *Glycolysis* $C_6H_{12}O_6$ (glucose) + 2 $NAD^+ \longrightarrow$ 2 $C_3H_4O_3$ (pyruvic acid) + 2 NADH
2. *Lactic Acid Creation* 2 $C_3H_4O_3$ (pyruvic acid) + 2 NADH $\longrightarrow$ 2 $C_3H_6O_3$ (lactic acid) + 2 NAD^+

Net Reaction: $C_6H_{12}O_6$ (glucose) $\longrightarrow$ 2 $C_3H_6O_3$ (lactic acid)

Fermentation in Other Organisms

In some plant cells and yeasts, fermentation produces CO_2 and ethanol. The conversion of pyruvate to acetaldehyde generates CO_2 and the conversion of acetaldehyde to ethanol regenerates NAD^+.

Anaerobic Respiration in Prokaryotes

In the field of prokaryotic metabolism, anaerobic respiration has a more specific meaning. In this case, anaerobic respiration is defined as a membrane bound biological process coupling the oxidation of electron donating substrates (e.g. sugars and other organic compounds, but also inorganic molecules like hydrogen, sulphide/sulphur, ammonia, metals or metal ions) to the reduction of suitable *external* electron acceptors other than molecular oxygen.

In contrast, in fermentation the oxidation of molecules is coupled to the reduction of an internally-generated electron acceptor, usually pyruvate. Hence, scientists who study prokaryotic physiology view anaerobic respiration and fermentation as distinct processes and therefore do not use the terms interchangeably.

In anaerobic respiration, as the electrons from the electron donor are transported down the electron transport chain to the terminal electron acceptor, protons are translocated over the membrane from "inside" to "outside", establishing a concentration gradient across the membrane which temporarily stores the energy released in the chemical reactions. This potential energy is then converted into ATP by the same enzyme used during aerobic respiration, ATP synthase.

Possible electron acceptors for anaerobic respiration are nitrate, nitrite, nitrous oxide, oxidised amines and nitro-compounds, fumarate, oxidised metal ions, sulphate, sulphur, sulphoxo-compounds, halogenated organic compounds, selenate, arsenate, bicarbonate or carbon dioxide (in acetogenesis and methanogenesis). All these types of anaerobic respiration are restricted to prokaryotic organisms.

Examples of anaerobic respiration:

Glucose + $3NO_3^- + 3H_2O \longrightarrow 6HCO_3^- + 3NH_4^+$,
$\Delta G^{0'} = -1796$ kJ

Glucose + $3SO_4^{2-} + 3H^+ \longrightarrow 6HCO_3^- + 3NH^-$,
$\Delta G^{0'} = -453$ kJ

Glucose + $12S + 12H_2O \longrightarrow 6HCO_3^- + 12HS^- +$
$18H^+$, $\Delta G^{0'} = -333$ kJ

All of these terminal electron acceptors are further upstream in the electron transport chain, compared to O_2. Consequently, anaerobic respiration is less effective than aerobic respiration. The $\Delta G^{0'}$ of aerobic respiration is -2844 kJ.

Stages of Photosynthesis

Photosynthesis uses light energy and carbon dioxide to make sugars like glucose. A general equation for photosynthesis is:

$$6\,CO_{2(gas)} + 12\,H_2O_{(liquid)} + photons \rightarrow C_6H_{12}O_{6(aqueous)} + 6\,O_{2(gas)} + 6\,H_2O_{(liquid)}$$

carbon dioxide + water + light energy →
glucose + oxygen + water

The equation is often presented in introductory chemistry texts in simplified form:

$$6\,CO_{2(gas)} + 6\,H_2O_{(liquid)} + photons \rightarrow C_6H_{12}O_{6(aqueous)} + 6\,O_{2(gas)}$$

Photosynthesis occurs in two stages. In the first phase *light-dependent reactions* or *photosynthetic reactions* (also called the *Light reactions*) capture the energy of light and use it to make high-energy molecules.

During the second phase, the *light-independent reactions* (also called the Calvin-Benson Cycle, and formerly known as the *Dark Reactions*) use the high-energy molecules to capture carbon dioxide (CO_2) and make the precursors of glucose.

In the light-dependent reactions one molecule of the pigment chlorophyll absorbs one photon and loses one electron. This electron excites pheophytin allowing the start of a flow of electrons down an electron transport chain that leads to the ultimate reduction of NADP into NADPH. In addition, it serves to create a proton gradient across the chloroplast membrane; its dissipation is used by ATP Synthase for the concomitant synthesis of ATP. The chlorophyll molecule regains the lost electron by taking one from a water molecule through a process called photolysis, that releases oxygen gas as a waste product.

In the Light-independent or dark reactions the enzyme Rubisco captures CO_2 from the atmosphere and in a process that requires the newly-formed NADPH, called the Calvin-Benson cycle releases three-carbon sugars which are later combined to form glucose. Photosynthesis may simply be defined as the conversion of light energy into chemical energy by living organisms. It is affected by its surroundings and the rate of photosynthesis is affected by the concentration of carbon dioxide, the intensity of light, and the temperature.

Light-dependent Reaction

The first stage of the photosynthetic system is the *light-dependent reaction*, which converts solar energy into chemical energy. The light dependent reaction produces oxygen gas and converts ADP and NADP+ into the energy carriers ATP and NADPH.

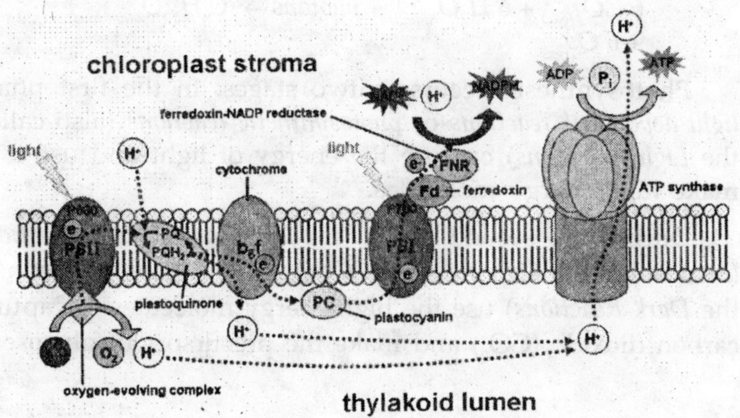

Electron Transport

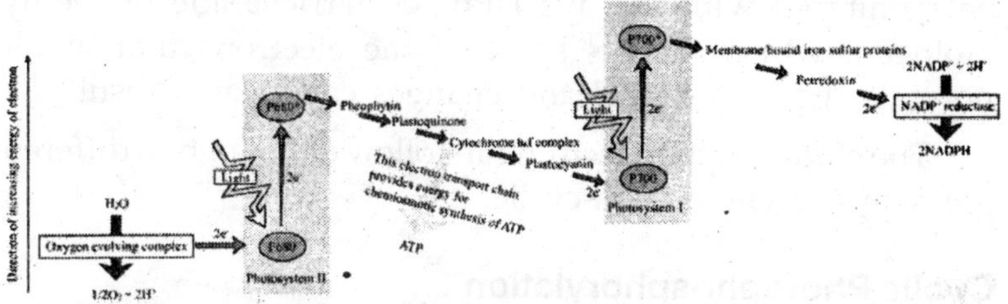

The process of synthesising ATP and NADPH is accomplished via the mechanism of an electron transport chain. This is a series of proteins embedded in a biological membrane that transfers high-energy electrons from one to another, accomplishing various activities along the way as the electron drops in energy level. When sunlight strikes a cluster of chlorophyll molecules, the molecule is dulled, and an electron is switched to a higher energy level of the molecule.

The excitation is transferred as an exciton from one antenna chlorophyll to another until it is captured by a primary reaction centre. It can be transferred from one molecule to another of the same kind of pigment, or from a carotenoid to chlorophyll, but not from chlorophyll to a carotenoid, because excitation of carotenoids carries more energy than that of chlorophyll. Only chlorophyll of the reaction centre is capable of transferring an electron to an electron acceptor (an intermediate, e.g., pheophytin in photosystem II and another chlorophyll molecule in Photosystem I).

Because the energy in light corresponds to its wavelength, the difference in excitation energy also allows the carotenoids to absorb light at wavelengths that chlorophyll does not absorb well. The P680 (photosystem II) and P700 (photosystem I) refer to reaction centre molecules of the two photosystems that receive excitons from other chlorophyll and accessory pigment molecules; each number 680 or 700 refers to the preferred wavelength of light absorbed, in the region of the spectrum, by the chlorophyll pigments at the respective reaction centres. These P680 and P700 molecules are in very low concentrations (1 molecule each per about 600 other chlorophyll molecules).

The rate of this stage of the light-dependent reactions can be monitored with the dye DPIP, or ferricyanide or methyl viologen, which accepts some of the electrons that would normally go to NADPH and changes colour as a result.

The chlorophyll's electron can follow either of two different pathways, cyclic or non-cyclic.

Cyclic Photophosphorylation

In *cyclic electron flow,* the electron begins in a pigment complex called photosystem I, passes from the primary acceptor to ferredoxin, then to a complex of two cytochromes (similar to those found in mitochondria), and then to plastocyanin before returning to chlorophyll. This transport chain produces a proton-motive force, pumping H^+ ions across the membrane; this produces a concentration gradient which can be used to power ATP synthase during chemiosmosis. This pathway is known as cyclic photophosphorylation, and it produces neither O_2 nor NADPH. In bacterial photosynthesis, a single photosystem is used, and therefore is involved in cyclic photophosphorylation.

Non-cyclic Photophosphorylation

The other pathway, non-cyclic photophosphorylation, is a two-stage process involving two different chlorophyll photosystems. First, a water molecule is broken down into $2H^+ + 1/2O_2 + 2e^-$. The two electrons from the water molecule are kept in photosystem II, while the $2H^+$ and $1/2O_2$ are left out for further use. Then a photon is absorbed by the chlorophyll core of photosystem II, exciting the two electrons which are transferred to the acceptor molecule. The deficit of electrons is replenished by taking electrons from another molecule of water. The electrons transfer from the primary acceptor to plastoquinone, then to plastocyanin, producing proton-motive force as with cyclic electron flow and driving ATP synthesis.

The photosystem II complex replaced its lost electrons from an external source, however, the two other electrons are not returned to photosystem II as they would in the analogous cyclic pathway. Instead, the still-excited electrons are transferred to a photosystem I complex, which boosts their energy level to a higher level using a second solar photon. The highly excited electrons are transferred to the acceptor molecule, but this time are passed on to an enzyme called Ferredoxin-NADP reductase/$NADP^+$ reductase, for short FNR, which uses them to catalyst the reaction (as shown):

$$NADP^+ + 2H^+ + 2e^- \rightarrow NADPH + H^+$$

This consumes the H^+ ions produced by the splitting of water, leading to a net production of $1/2O_2$, ATP, and $NADPH+H^+$ with the consumption of solar photons and water.

The concentration of NADPH in the chloroplast may help regulate which pathway electrons take through the light reactions. When the chloroplast runs low on ATP for the Calvin cycle, NADPH will accumulate and the plant may shift from non-cyclic to cyclic electron flow.

Steps

It is important to note that both photosystems are almost simultaneously excited; thus, both photosystems begin functioning at almost the same time.

1. Light strikes photosystem II and the energy is absorbed and passed along until it reaches P680 chlorophyll.
2. The excited electron is passed to the primary electron acceptor. Photolysis in the thylakoid takes the electrons from water and replaces the P680 electrons that were passed to the primary electron acceptor. (O_2 is released as a waste product)
3. The electrons are passed to photosystem I via the electron transport chain (ETC) and in the process used to pump protons across the thylakoid membrane into the lumen.
4. The stored energy in the proton gradient is used to produce ATP which is used later in the Calvin-Benson Cycle.

5. P700 chlorophyll then uses light to excite the electron to its second primary acceptor.
6. The electron is sent down another ETC and used to reduce $NADP^+$ to NADPH.
7. The NADPH is then used later in the Calvin-Benson Cycle.

Calvin Cycle

The *Calvin cycle* (or *Calvin-Benson cycle* or carbon fixation) is a series of bio-chemical reactions that takes place in the stroma of chloroplasts in photosynthetic organisms. It was discovered by Melvin Calvin and Andrew Benson at the University of California, Berkeley with James Bassham also contributing. It is one of the light-independent reactions or dark reactions.

During photosynthesis, light energy is used to generate chemical free energy, stored in glucose. The light-independent Calvin cycle, also (misleadingly) known as the "dark reaction" or "dark stage", uses the energy from short-lived electronically-excited carriers to convert carbon dioxide and water into organic compounds that can be used by the organism (and by animals which feed on it). This set of reactions is also called *carbon fixation*. The key enzyme of the cycle is called Rubisco. In the following equations, the chemical species (phosphates and carboxylic acids) exist in equilibria among their various ionised states as governed by the pH.

The enzymes in the Calvin cycle are functionally equivalent to many enzymes used in other metabolic pathways such as glycolysis and gluconeogenesis, but they are to be found in the chloroplast stroma instead of the cell cytoplasm, separating the reactions. They are activated in the light (which is why the name "dark reaction" is misleading), and also by products of the light-dependent reaction. These regulatory functions prevent the Calvin cycle from operating in reverse to respiration, which would create a continuous cycle of carbon dioxide being reduced to carbohydrates, and carbohydrates being respired to carbon dioxide. Energy (in the form of ATP) would be wasted in carrying out these reactions that have no net productivity.

The sum of reactions in the Calvin cycle is the following:

$$3\ CO_2 + 6\ NADPH + 5\ H_2O + 9\ ATP \rightarrow C_3H_5O_3\text{-}PO_3^{2-} + 2\ H^+ + 6\ NADP^+ + 9\ ADP + 8\ P_i$$

It should be noted that hexose (six carbon) sugars are not a product of the Calvin cycle. Although many texts list a product of photosynthesis as $C_6H_{12}O_6$, this is mainly a convenience to counter the equations of respiration, where six-carbon sugars are oxidised in mitochondria. The carbohydrate products of the Calvin Cycle are three-carbon sugar phosphate molecules, or "triose phosphates," specifically, glyceraldehyde-3-phosphate.

Steps of the Calvin Cycle

- The enzyme Rubisco catalyses the carboxylation of Ribulose-1,5-bisphosphate, a 5 carbon compound, by carbon dioxide (a total of 6 carbons). Two molecules of glycerate 3-phosphate, a 3-carbon compound, are created. (also: 3-phosphoglycerate, 3-phosphoglyceric acid, 3PGA)
- The enzyme phosphoglycerate kinase catalyses the phosphorylation of 3PGA by ATP (which was produced in the light-dependent stage). 1,3-bisphosphoglycerate (glycerate-1,3-bisphosphate) and ADP are the products. (However, note that two PGAs are produced for every CO_2 that enters the cycle, so this step happens twice.)
- The enzyme G3P dehydrogenase catalyses the reduction of 1,3 BPGA by NADPH (which was another product of the light-dependent stage). Glyceraldehyde 3-phosphate (also G3P, GP) is produced, and the NADPH itself was oxidised and hence becomes $NADP^+$.

(Simplified versions of the Calvin cycle integrate the

remaining steps, except for the last one, into one general step- the regeneration of RuBP - also, one G3P would exit here.)

- Triose phosphate isomerase converts some G3P reversibly into dihydroxyacetone phosphate (DHAP), also a 3-carbon molecule.
- Aldolase and Fructose-1,6-bisphosphatase convert some of these two into fructose-6-phosphate (6C). A phosphate ion is lost to ADP.

Up to this point, as per the overall equation, 6 carbon dioxide molecules would have been converted, with the use of 6 RuBP, 12 ATP and 12 NADPH, to 12 G3P molecules. One F6P, (= 2 G3P) then exits the cycle, while 10 of these G3P molecules continue, giving a ratio of 1:5 G3P. Obviously, the ratio of carbon dioxide entering the cycle to RuBP already present is also 1:5.

- F6P is then combined with another G3P (total 9C) and then cleaved into xylulose-5-phosphate (X5P) and erythrose-4-phosphate by transketolase.
- E4P and DHAP are converted into sedoheptulose-7-phosphate (7C) by S1,7BPase. A phosphate ion is lost to ADP.
- S7P is then combined with another G3P (total 10C) and then cleaved into another X5P and ribose-5-phosphate (R5P) again by transketolase.
- X5P is converted into ribulose-5-phosphate (Ru5P, RuP) by phosphopentose epimerase. R5P is also converted into RuP by ribose isomerase.
- Finally, phosphoribulokinase phosphorylates RuP into RuBP, ribulose-1,5-bisphosphate, completing the Calvin *cycle*. This requires the input of one ATP.

All the G3P produced earlier is converted into RuBP (5C), so 10 G3Ps (30C, 10 phosphates) were needed to produce 6 RuBPs (30C, 6 phosphates). 6 ATPs were also needed in the last step, giving a total of 18 ATPs used up per 6 CO_2s. However, four phosphate ions are lost and these also form ATP. The energy in those ATPs is used to drive some of the reactions. At high temperatures, Rubisco will react with O_2 instead of CO_2 in *photorespiration*. This turns RuBP into 3PGA and 2-phosphoglycolate, a 2-carbon molecule which can be converted into 3PGA, some of which will exit the Calvin cycle. However, if this continues the RuBP will eventually be depleted, which slows down the cycle if electrons are entering from the light-dependent reaction too quickly.

Products of the Calvin Cycle

The immediate product of the Calvin cycle is glyceraldehyde-3-phosphate (G3P). Two G3P molecules (or one F6P molecule) that have exited the cycle are used to make larger carbohydrates. In simplified versions of the Calvin cycle they may be converted to F6P or F5P after exit, but this conversion is also part of the cycle.

Hexose isomerase converts about half of the F6P molecules into glucose-6-phosphate. These are dephosphorylated and the glucose can be used to form starch, which is stored in, for example, potatoes, or cellulose used to build up cell walls. Other glucose, with fructose, forms sucrose, the plant sugar.

In plants

Most plants are photoautotrophs, which means that they are able to synthesise food directly from inorganic compounds using light energy - for example from the sun, instead of eating other organisms or relying on nutrients derived from them. This is distinct from chemoautotrophs that do *not* depend on light energy, but use energy from inorganic compounds.

The energy for photosynthesis ultimately comes from absorbed photons and involves a reducing agent, which is water in the case of plants, releasing oxygen as a waste product.

The light energy is converted to chemical energy (known as light-dependent reactions), in the form of ATP and NADPH, which are used for synthetic reactions in photoautotrophs. The overall equation for the light-dependent reactions under the conditions of non-cyclic electron flow in green plants is:

$$2\ H_2O + 2\ NADP^+ + 2\ ADP + 2\ P_i\ (+\ light) \rightarrow 2\ NADPH + 2\ H^+ + 2\ ATP + O_2$$

Most notably plants use the chemical energy to fix carbon dioxide into carbohydrates and other organic compounds through light-independent reactions. The overall equation for carbon fixation (sometimes referred to as carbon reduction) in green plants is:

$$3\ CO_2 + 9\ ATP + 6\ NADPH + 6\ H^+ \rightarrow C_3H_6O_3\text{-phosphate} + 9\ ADP + 8\ P_i + 6\ NADP^+ + 3\ H_2O$$

More specifically, carbon fixation produces an intermediate product, which is then converted to the final hexose carbohydrate products. These carbohydrate products are then variously used to form other organic compounds, such as the building material cellulose, as precursors for lipid and amino acid biosynthesis or as a fuel in cellular respiration. The latter not only occurs in plants, but also in animals when the energy from plants get passed through a food chain. Organisms dependent on photosynthetic and chemosynthetic organisms are called heterotrophs. In general outline, cellular respiration is the opposite of photosynthesis: glucose and other compounds are oxidised to produce carbon dioxide, water, and chemical energy. However, both processes actually take place through a different sequence of reactions and in different cellular compartments.

Plants absorb light primarily using the pigment chlorophyll, which is the reason that most plants have a green colour. The function of chlorophyll is often supported by other accessory pigments such as carotenes and xanthophylls. Both chlorophyll and accessory pigments are contained in organelles (compartments within the cell) called Chloroplasts. Although all cells in the green parts of a plant have chloroplasts, most of the energy is captured in the leâves. The cells in the interior tissues of a leaf, called the mesophyll, contain about half a million chloroplasts for every square millimetre of leaf. The surface of the leaf is uniformly coated with a water-resistant waxy cuticle that protects the leaf from excessive evaporation of water and decreases the absorption of ultraviolet or blue light to reduce heating. The transparent epidermis layer allows light to pass through to the palisade mesophyll cells where most of the photosynthesis takes place.

In Algae and Bacteria

Algae is a range from multicellular forms like kelp, to microscopic, single-celled organisms. Although they are not as complex as land plants, photosynthesis takes place biochemically the same way. Very much like plants, algae have chloroplasts and chlorophyll, but various accessory pigments are present in some algae such as phycoerythrin in red algae (rhodophytes), resulting in a wide variety of colours. All algae produce oxygen, and many are autotrophic. However, some are heterotrophic, relying on materials produced by other organisms. For example, in coral reefs, there is a symbiotic relationship between zooxanthellae and the coral polyps.

Photosynthetic bacteria do not have chloroplasts (or any membrane-bound organelles). Instead, photosynthesis takes place directly within the cell. Cyanobacteria contain thylakoid membranes very similar to those in chloroplasts and are the only prokaryotes that perform oxygen-generating photosynthesis. In fact chloroplasts are now considered to have evolved from an endosymbiotic bacterium, which was also an ancestor of and later gave rise to cyanobacterium. The other photosynthetic bacteria have a variety of different pigments, called bacteriochlorophylls, and do not produce oxygen. Some bacteria, such as *Chromatium*, oxidise hydrogen sulphide instead of water for photosynthesis, producing sulphur as waste.

Photosynthetic Organisms

There are three main classes of organisms which use light energy. The first two show many similarities. The Halobacteria are quite different and are often not considered as truly photosynthetic. They.are dealt with separately.

(A) Cyanobacteria and green plants (chloroplasts probably evolved from cyanobacteria-like endosymbionts). Oxygenic photosynthesis - O_2 is produced and photosystems I and II are both present. Fix CO_2.

(B) *Photosynthetic Eubacteria:* Anaerobic photosynthesis-no O_2 is produced and only photosystem I is present. Sometimes fix CO_2 but need an external reductant such as H_2S or succinate.

(C) *Halobacteria:* No chlorophyll, no photosynthetic electron transport. Possess a light-driven proton pump which produces energy but no reducing power. Need source of organic carbon. Cannot Fix CO_2.

Main Stages of Photosynthesis

(I) Light is absorbed and used to generate proton motive force and reducing power (as NADH or NADPH).

(II) Proton motive force is used to generate ATP.

Once the PMF has been generated by the photosystems, the mechanism of ATP synthesis is essentially the same as in respiration. An ATP synthetase is found in the photosynthetic membranes works just like the one in respiratory membranes.

III) ATP and NAD(P)H are used to fix carbon dioxide.

The supply of ATP and NAD(P)H do not have to come from light energy. Autotrophic bacteria which use inorganic reactions as a source of energy can often also fix CO_2.

Trapping Light Energy

Light is absorbed by chlorophyll located in the photosynthetic membranes. When light is absorbed, an electron is released from the chlorophyll and travels down an electron transport chain. However, only a few special chlorophyll molecules in the reaction centre actually release electrons. Most of the chlorophyll is found in the antenna which absorbs light energy and funnels it to the reaction centre chlorophyll, where electron release occurs. For every reaction centre chlorophyll there may be several hundred antenna molecules. The antenna and reaction centre chlorophylls are chemically identical, they differ due to the proteins which bind them.

A typical antenna complex consists of a small to medium size protein which carries several each of chlorophyll *a* and chlorophyll *b* and a couple of carotenoids. Precise details vary between organisms. The two types of chlorophyll absorb light of different wavelength. The carotenoids absorb well in regions of the spectrum where chlorophyll absorbs poorly. In addition, carotenoids help protect against high light intensity. Bacteriochlorophyll (Bchl) is slightly different chemically from plant chlorophyll and absorbs light of longer wavelength (lower energy).

The reaction centre consists of a cluster of proteins plus a pair of chlorophyll molecules and some electron carriers. Plants and cyanobacteria have two distinct reaction centres which operate in series whereas anaerobic photosynthetic bacteria have only a single reaction centre. The electron ejected from the reaction centre chlorophyll travels along a series of electron carriers and proton motive force is generated. In anaerobic bacteria, this electron cycles back to the reaction centre. In cyanobacteria and chloroplasts, the electron continues on to reduce NADP after a second boost of energy from the second photosystem.

Purple Photosynthetic Bacteria

We will consider first the simpler anaerobic photosynthesisers which only possess one type of reaction centre. The details of the reaction centre complex varies among different bacteria. X-ray crystallography has been used to analyse the reaction centre from *Rhodopseudomonas viridis*.

The components of the reaction centre are:

Polypeptide chains (L, M, H & C) four
bacteriochlorophyll *b* (Bchl) four (shared by L & M)

Bacteriopheophytin (Bphe) two (shared by L & M)

Ubiquinone (UQ) two (shared by L & M)

Ferrous iron (non-haem) one (shared by L & M)

Haem four (on protein C)

When chlorophyll absorbs a photon, an electron is excited to a higher-energy orbital. This energy may be released by any of:

(a) Emit a photon, i.e. fluorescence. Some energy is lost so the wavelength of the emitted photon is slightly longer than that of the absorbed photon.

(b) Emit heat.

(c) Transfer energy to neighbouring molecule, as when antenna chlorophyll transfers energy to the reaction centre. The energy may actually travel via many antenna chlorophylls before reaching P960.

(d) Eject an excited electron, as occurs in the reaction centre.

Sequence of Events:

a. Light is absorbed by Bchl in the antenna.
b. Energy is transferred to P960 which is a pair of special Bchl *b* molecules in the reaction centre. (P960 means pigment with absorption maximum at 960nm)
c. P960 becomes excited. The excited state, P960*, lasts less than one picosecond (pico = 10-15).
d. P960 loses an electron to one of the other Bchl *b* molecules in the reaction centre.
e. The electron is rapidly passed on to Bphe. Bphe is a chlorophyll derivative where Mg_2+ is replaced by 2H+.

The transient existence of the biradical (Bchl)2+....Bphe- is detectable by its spectrum.

f. The electron is transferred from Bphe to firmly bound ubiquinone, UQA, to give the half-reduced form (the anionic semiquinone). UQA is never fully reduced and never gains protons to become UQH_2.

g. The electron goes from UQA to the loosely bound ubiquinone, UQB. UQB waits for a second electron and also picks up two protons from water so becoming UQH_2.

h. $UQBH_2$ moves out of the reaction centre, travels through the lipid bilayer, and transfers its electrons to the cytochrome *bc*1 complex.

i. Cytochrome *bc*1 transfers electrons to cytochrome *c*2 which is a soluble periplasmic protein.

j. The electrons are transferred from cytochrome *c*2 to reaction centre protein C and travel successively via its four haemes back to P960+.

Generation of Proton Motive Force

Two protons are translocated across the membrane, from the cytoplasm to the periplasm, for each electron which goes around this loop. This creates the proton motive force which is used to make ATP. This is known as cyclic photophosphorylation by analogy to oxidative phosphorylation.

Generation of Reducing Power

Higher plants produce NADPH but purple bacteria often generate NADH during photosynthesis. They do so by reversed electron transport. Electrons from P960 flow to ubiquinone and then, instead of going via cytochrome *bc*1, they go to NADH dehydrogenase which reduces NAD+ to NADH. The NADH dehydrogenase is driven in reverse by the proton gradient (i.e. electrons are travelling uphill, energetically, and hence energy input is required).

Since the electrons are not returned to the P960 they must be replaced somehow. In practice purple bacteria need an external supply of electrons. For purple sulphur bacteria, the electron donor is H_2S or thiosulphate. Non-sulphur bacteria use organic donors, e.g. malate or succinate.

Respiration in Purple Bacteria

Many purple bacteria can also respire aerobically in the dark. They use NAD for both the respiratory system and the photosynthetic system, depending on the circumstances. (Chloroplasts do not respire and use NADPH for only one purpose - fixing CO_2) Oxygen represses the synthesis of chlorophyll and carotenoids but the rest of the electron transport system remains and cytochrome *bc*1 transfers electrons to cytochrome *a*/*a*3 which is induced by O_2.

The central part of the electron transport chain is shared by respiration and photosynthesis. During respiration the PMF is generated by ejecting protons across the membrane into the periplasmic space. There are three sites - between NADH dehydrogenase and ubiquinone, at cytochrome *bc*1 and between cytochrome *a*/*a*3 and O_2.

Evolutionary Relationships

The order in which the ability to generate energy evolved is generally considered to be:

1. Anaerobic fermentation: ATP converted to PMF
2. Anaerobic photosynthesis: light converted to PMF
3. Oxygenic phtosynthesis: light converted to PMF and oxygen evolved
4. Respiration: reverse electron transport chain uses O_2

Ribosomal RNA sequence homology indicates that chloroplasts are related to gram-negative bacteria. Chloroplasts are prokaryotic in structure and probably derived from an ancestor related to cyanobacteria which was first symbiotic inside eukaryotic cells and finally became an organelle. Cyanobacteria lack chlorophyll *b* and instead possess phycobilins whereas green plants do not. However, *Prochloron* and its relatives are prokaryotic oxygenic photosynthesisers without phycobilins but which do have chlorophyll *b*. Hence they are classified as prochlorophytes not cyanobacteria, and may be closer to the chloroplast ancestor.

Bacterial Metabolism and Photosynthesis

One of the most staggering aspects of bacteriology is the diversity of metabolism these organisms display. There is no known naturally-occurring carbon-containing molecule which cannot be metabolised by at least one species. Originally, two categories of metabolism were recognised:

- *Autotrophs:* Organisms which use only inorganic nutrients, e.g. plants.
- *Heterotrophs:* Organisms which require organic carbon sources, e.g. animals.

However, this simple classification cannot accurately describe the metabolic diversity of bacteria. In order to accomplish this, bacterial metabolic types are usually defined on basis of both the energy and carbon source used:

Autotrophs

Energy Source	*Carbon Source*	*Name*	*Example*
Light	Inorganic	Photoautotroph	Most photosynthetic bacteria, e.g. *Chromatium* (anaerobic), Cyanobacteria (aerobic)
Inorganic	Inorganic	Chemolithotrophic autotroph (chemoautotroph)	*Nitrobacter*
Organic	Inorganic	Chemoorganotrophic autotroph	*Pseudomonas oxalaticus*

Heterotrophs

Energy Source	*Carbon Source*	*Name*	*Example*
Light	Organic	Photoheterotroph	Purple and green photosynthetic bacteria, e.g. *Rhodospirillum*
Inorganic	Organic	Chemolithotrophic heterotroph ("mixotroph")	*Desulphovibrio*
Organic	Organic	(Chemo)heterotroph	*E. coli*

The order *Rhodospirillales* contains three families of phototrophic bacteria:

Rhodospirillaceae: Purple non-sulphur bacteria, e.g. *Rhodospirillum.* Cells of *Rhodospirillum* are helical, ~1 µm in diameter and variable in length. These cells contain bacteriochlorophyll a or b located on specialised membranes continuous with the cytoplasmic membrane. They are not able to use elemental sulphur as electron donor and typically use an organic electron donor, such as succinate or malate, but can also use hydrogen gas.

Chromatiaceae: Purple sulphur bacteria, e.g. *Chromatium.* These are short, Gram-negative rods, ~1 µm in diameter and 3-4 µm long. They are able to use sulphur and sulphide as the sole photosynthetic electron donor and sulphide can be oxidised to sulphate. These bacteria use an inorganic sulphur compound, such as hydrogen sulphide as an electron donor. Purple sulphur bacteria must fix CO_2 to live, whereas non-sulphur purple bacteria can grow aerobically in the dark by respiration on an organic carbon source.

The metabolism of living organisms as they evolved on earth constantly dissipates energy in a non-reusable form. To sustain life this energy has to be replenished. The only external energy source is the sun, so life has to cope with utilising light to fuel all metabolic demands.

Photons emanated by the sun carry energy depending on the colour of the light. At a wavelength of 800 µm this amounts to 35.7 kcal/mol, which compares favourably with the hydrolysis energy of, e.g. phosphoenolpyruvate (14.8 kcal/mol) or adenosine triphosphate (7.3 kcal/mol). So all we need is a machinery to catch the photon's energy and store it in the form of say a phosphoester bond. Nature accomplishes this task with the aid of multienzyme complexes organised in cellular membranes. The simplest systems are found in purple photosynthetic bacteria. The cytoplasmic membranes contain photosynthetic units (PSU) which are made up of reaction centres (RC) and light harvesting complexes (LH I and LH II).

Light drives the reduction of quinones by the PSU and the quinones are reoxidised by a cytochrome complex (thus turning the process into a cycle). Turning the quinone cycle expels protons from the cytoplasm to the periplasm. The electrochemical gradient thus generated is used to drive ATP synthesis by the membrane bound F_0F_1-ATPase.

Besides quinones other ligands are used by the photosynthetic enzymes: chlorophylls are used as photon acceptors, carotenoid mediate the transfer of electrons to the quinones, and haemes are used by the cytochromes to recycle electrons. The exact type of ligands varies according to the bacteria as does the arrangement of the protein subunits in the different complexes. Several structures were resolved to the atomic level giving insight to the mechanisms of energy conversion. The supermolecular organisation is so far known by cryo-electron microscopy.

In the frame to the right the chromophores of the reaction centre of *Rhodopseudomonas sphaeroides* are to be seen. The primary chlorophyll (type a) is shown in yellow. The electrons are passed via another bacteriochlorophyll a (green) and a bacteriopheophytin (light blue) to ubiquinone-10 (red). Magnesium atoms are symbolised in dark green, a single iron ion in orange.

The primary chlorophyll in the reaction centre is rather a small target to be hit by photons. To improve the absorption cross-section, RCs are surrounded by antenna proteins harbouring more chlorophyll as well as pigments absorbing at other wavelengths thus extending the efficiency even further. These proteins are arranged in circular patterns around the RCs within the bacterial cytoplasmic membrane. The diameter of these complexes extends to ca. 10 μm, they are termed light harvesting complex I (LH I). Most bacteria contain additional photosensitive complexes (LH II). An example of the chromophore arrangement of a LH II is shown at the top of this page. Below two different complexes are shown with their additional carotenoid chromophores and the supporting protein structure. In the bacterial membranes several LH II complexes are situated close enough to LH I to mediate fast energy transfer via LH I to the RC.

Chloroflexus aurantiacus is a photosynthetic bacterium isolated from hot springs, belonging to the green non-sulphur bacteria. This organism is thermophilic and can grow at temperatures from 35 °C to 70 °C. *Chloroflexus aurantiacus* can survive in the dark if oxygen is available. When grown in the dark, *Chloroflexus aurantiacus* has a dark orange colour. When grown in sunlight it is dark green. The individual bacteria tend to form filamentous colonies enclosed in sheaths, which are known as trichomes.

As a genus, *Chloroflexus* spp. are gram negative filamentous anoxygenic phototrophic (FAP) organisms that utilise type II photosynthetic reaction centres containing bacteriochlorophyll *a* similar to the purple bacteria, and light-harvesting chlorosomes containing bacteriochlorophyll *c* similar to green sulphur bacteria of the *Chlorobi*.

As the name implies, these anoxygenic phototrophs do not produce oxygen as a by-product of photosynthesis, in contrast to oxygenic phototrophs such as cyanobacteria, algae, and higher plants. While oxygenic phototrophs use water as an electron donor for phototrophy, *Chloroflexus* uses reduced sulphur compounds such as hydrogen sulphide, thiosulphate, or elemental sulphur. This belies their antiquated name *green non-sulphur bacteria*, however *Chloroflexus* spp. can also utilise hydrogen (H_2) as a source of electrons.

Chloroflexus aurantiacus is thought to grow photoheterotrophically in nature, but it has the capability of fixing inorganic carbon through photoautotrophic growth. Instead of using the Calvin-Benson-Bassham Cycle typical of plants, *Chloroflexus aurantiacus* has been demonstrated to use a novel autotrophic pathway known as the 3-hydroxypropionate pathway.

The complete electron transport chain for *Chloroflexus* spp.

is not yet known. Particularly, *Chloroflexus aurantiacus* has not been demonstrated to have a cytochrome bc_1 complex, and may use different proteins to reduce cytochrome *c*.

One of the main reasons for interest in *Chloroflexus aurantiacus* is in the study of the evolution of photosynthesis. As terrestrial mammals, we are most familiar with photosynthetic plants such as trees. However, photosynthetic eukaryotes are a relatively recent evolutionary development. Photosynthesis by eukaryotic organisms can be traced back to endosymbiotic events in which non-photosynthetic eukaryotes internalised photosynthetic organisms. The chloroplasts of trees still retain their own DNA as a molecular remnant that indicated their origin as photosynthetic bacteria.

Leaf, Mesophyll and Photosynthesis

In botany, a leaf is an above-ground plant organ specialised for photosynthesis. For this purpose, a leaf is typically flat (laminar) and thin, to expose the cells containing chloroplast (chlorenchyma tissue, a type of parenchyma) to light over a broad area, and to allow light to penetrate fully into the tissues. Leaves are also the sites in most plants where respiration, transpiration, and guttation take place. Leaves can store food and water, and are modified in some plants for other purposes. The comparable structures of ferns are correctly referred to as fronds. Leaves are prominent in the human diet as leaf vegetables. Structurally complete leaf of an angiosperm consists of a petiole (leaf stem), a *lamina* (leaf blade), and stipules (small processes located to either side of the base of the petiole). The point at which the petiole attaches to the stem is called the leaf axil. Not every species produces leaves with all of these structural parts. In some species, paired stipules are not obvious or are absent altogether. A petiole may be absent, or the blade may not be laminar (flattened).

A leaf is considered to be a plant organ, typically consisting of the following tissues:

1. An *epidermis* that covers the upper and lower surfaces
2. An interior *chlorenchyma* called the mesophyll
3. An arrangement of veins (the vascular tissue).

The C_4 pathway was discovered by M. D. Hatch and C. R. Slack, two Australian researchers, in 1966, so it is sometimes called the Hatch-Slack pathway.

Epidermis

The epidermis is the outer multilayered group of cells covering the leaf. It forms the boundary between the plant and the external world. The epidermis serves several functions: protection against water loss, regulation of gas exchange, secretion of metabolic compounds, and (in some species) absorption of water. Most leaves show dorsoventral anatomy: the upper (adaxial) and lower (abaxial) surfaces have somewhat different construction and may serve different functions.

The epidermis is usually transparent (epidermal cells lack chloroplasts) and coated on the outer side with a waxy cuticle that prevents water loss. The cuticle may be thinner on the lower epidermis than on the upper epidermis, and is thicker on leaves from dry climates as compared with those from wet climates.

The epidermis tissue includes several differentiated cell types: epidermal cells, guard cells, subsidiary cells, and epidermal hairs (trichomes). The epidermal cells are the most numerous, largest, and least specialised. These are typically more elongated in the leaves of monocots than in those of dicots.

The epidermis is covered with pores called *stomata*, part of a stoma complex consisting of a pore surrounded on each side by chloroplast-containing guard cells, and two to four subsidiary cells that lack chloroplasts. The stoma complex regulates the exchange of gases and water vapour between the outside air and the interior of the leaf. Typically, the stomata are more numerous over the abaxial (lower) epidermis than the adaxial (upper) epidermis.

Mesophyll

Most of the interior of the leaf between the upper and lower layers of epidermis is a *parenchyma* (ground tissue) or *chlorenchyma* tissue called the *mesophyll* (Greek for "middle leaf"). This assimilation tissue is the primary location of photosynthesis in the plant. The products of photosynthesis are called "assimilates"

Mesophyll cells constitute the main body of a leaf, occurring between the upper and lower epidermis. Typically, the leaves of temperate-zone plants have two layers of mesophyll cells, the palisade mesophyll on the adaxial (upper) side, and the spongy mesophyll on the abaxial (lower) side. The palisade mesophyll is a layer of densely packed, columnar cells which contain many chloroplasts.

This layer is responsible for most of the photosynthesis of leaves. The spongy mesophyll is composed of large, often odd-shaped, photosynthetic cells separated from one another by large, intercellular spaces. The intercellular spaces apparently facilitate the exchange of photosynthetic gases.

In ferns and most flowering plants the mesophyll is divided into two layers:

- An upper *palisade layer* of tightly packed, vertically elongated cells, one to two cells thick, directly beneath the adaxial epidermis. Its cells contain many more chloroplasts than the spongy layer. These long cylindrical cells are regularly arranged in one to five rows. Cylindrical cells, with the chloroplasts close to the walls of the cell, can take optimal advantage of light. The slight separation of the cells provides maximum absorption of carbon dioxide. This separation must be minimal to afford capillary action for water distribution. In order to adapt to their different environment (such as sun or shade), plants had to adapt this structure to obtain optimal result. Sun leaves have a multilayered palisade layer, while shade leaves or older leaves closer to the soil, are single-layered.

- Beneath the palisade layer is the *spongy layer*. The cells of the spongy layer are more rounded and not so tightly packed. There are large intercellular air spaces. These cells contain fewer chloroplasts than those of the palisade layer.

The pores or *stomata* of the epidermis open into substomatal chambers, connecting to air spaces between the spongy layer cells.

These two different layers of the mesophyll are absent in many aquatic and marsh plants. Even an epidermis and a mesophyll may be lacking. Instead for their gaseous exchanges they use a homogeneous *aerenchyma* (thin-walled cells separated by large gas-filled spaces). Their stomata are situated at the upper surface.

Leaves are normally green in colour, which comes from chlorophyll found in plastids in the chlorenchyma cells. Plants that lack chlorophyll cannot photosynthesise.

Mesophyll cells are specialised for photosynthesis. These cells in the middle of the leaf contain many chloroplasts, the organelles that perform photosynthesis. As you demonstrate your knowledge of plant cell structure by placing the labels in the appropriate locations, remember that gas exchange (oxygen and carbon dioxide) takes place within the chloroplasts, and the food produced by the chloroplasts must move out of the cells to other parts of the plant.

The palisade mesophyll tissue is a layer of cells in a leaf. These cells all contain numerous chloroplasts, which contain the green chemical required for photosynthesis. At the top of the slide you can also see some epidermal cells. These do not contain chloroplasts. Their job is to keep water in the leaf and stop bacteria and fungi from getting in.

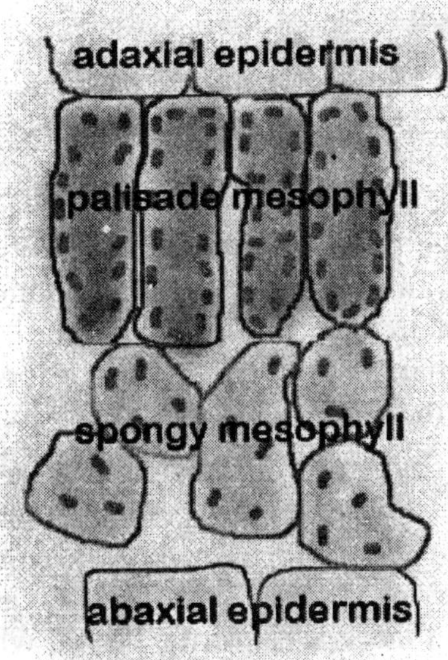

A simple cross-section through parenchyma (palisade and spongy) of the mesophyll of a leaf

The mesophyll of the leaf consists of parenchyma tissue between the abaxial and adaxial epidermis of the leaf. The mesophyll cells contain chloroplasts and it is in these cells that photosynthesis takes place. In many tree species, two distinct layers of mesophyll cells may be recognised, the palisade mesophyll and the spongy mesophyll.

Palisade Mesophyll

The palisade mesophyll consists of one or several layers of elongated, narrow parenchyma cells with their long axes at right angles to the axis of the leaf and are situated under the adaxial epidermis. Chloroplasts are especially concentrated in the palisade mesophyll and it is in these cells that much of the photosynthesis in a tree takes place.

Spongy Mesophyll

The spongy mesophyll consists of irregularly shaped parenchyma cells which are located between the palisade mesophyll and the abaxial epidermis. Spongy mesophyll cells contain less chloroplast than the palisade mesophyll cells but photosynthesis take place in these cells as well. There are many intercellular air spaces between the spongy mesophyll cells which are interconnected and communicate with the stomata of the abaxial epidermis. This allows the food-producing cells of the leaf (the mesophyll) to access the gases (carbon dioxide - CO_2 and oxygen - O_2) which they need for photosynthesis and respiration.

Through the leaf tissues run numerous interconnected veins or vascular bundles of varying sizes. These veins transport water and dissolved mineral salts into the leaf which are used for the production of organic substances during photosynthesis and related chemical pathways, which take place in the leaf. They are also responsible for transporting these substances away from the leaves to the other parts of the tree where they will be used. The larger veins of the leaf provide the mesophyll of the leaf with support.

The larger veins form the midrib and lateral veins of the leaf but there are also numerous small veins which can only be seen with the aid of a microscope. The larger veins have xylem with vessels and tracheids, and phloem with sieve tubes and companion cells.

In these larger veins, the xylem is orientated towards the adaxial epidermis. The small veins have fewer vascular cells and no supporting cells. The veins are surrounded by parenchyma cells, through which water and dissolved organic and inorganic substances must pass when entering or leaving the vein. The bundle sheath may or may not contain chloroplasts. The vascular tissue of the leaf is continuous with that of the stem.

Mesophyll structure has been associated with the photosynthetic performance of leaves via the regulation of internal light and CO_2 profiles. Differences in mesophyll structure and chlorophyll distribution within three ontogenetically different leaf types of *Eucalyptus globulus* ssp. *globulus* were investigated. Juvenile leaves are blue-grey in colour, dorsoventral (adaxial palisade layer only), hypostomatous, and approximately horizontal in orientation. In contrast, adult leaves are dark green in colour, isobilateral (adaxial and abaxial palisade), amphistomatous, and nearly vertical in orientation.

The transitional leaf type has structural features that appear intermediate between the juvenile and adult leaves. The ratio of mesophyll cell surface area, per unit leaf surface area (A_{mes}/A) of juvenile leaves was maximum at the base of a single, adaxial palisade layer and declined through the spongy mesophyll. Chlorophyll $a + b$ content showed a coincident pattern, while the chlorophyll $a{:}b$ ratio declined linearly from the adaxial to abaxial epidermis. In comparison, the mesophyll of adult leaves had a bimodal distribution of A_{mes}/A, with maxima occurring beneath both the adaxial and abaxial surfaces within the first layer of multiple palisade layers. The distribution of chlorophyll $a + b$ content had a similar pattern, although the maximum ratio of chlorophyll $a{:}b$ occurred immediately beneath the adaxial and abaxial epidermis. The matching distributions of A_{mes}/A and chlorophyll provide further evidence that mesophyll structure may act to influence photosynthetic performance. These changes in internal leaf structure at different life stages of *E. globulus* may be an adaptation for increased xeromorphy under increasing light exposure experienced from the seedling to adult tree, similar to the characteristics reported for different species according to sunlight exposure and water availability within their native habitats.

Container-grown black spruce (*Picea mariana* (Mill.) B.S.P.) seedlings were planted in trays containing a sand and peat mixture, and placed in a climate-controlled greenhouse. One group of seedlings was kept well-watered, and another group was subjected to three cycles of drought. Gas exchange analysis showed that mesophyll photosynthetic function was largely unimpaired by drought. In contrast, stomatal conductance was sensitive to drought, although it became less sensitive with each drought cycle. Both stomatal and mesophyll conductances increased with time in control and drought-stressed seedlings, but mesophyll conductance increased with time more rapidly than did stomatal conductance. Limitation of photosynthetic rate was dominated by the mesophyll. In control seedlings, relative stomatal limitation increased from 6 to 16 per cent by the end of the experiment. In drought-stressed seedlings, relative stomatal limitation of photosynthesis reached 40 per cent during the first drought, but decreased to near control values immediately after rewatering. Because the third, most severe drought had only a minor effect on stomatal conductance, relative stomatal limitation of photosynthesis was similar to that in control seedlings by the end of the experiment. Inhibition of ontogenetic change during drought stress may be responsible for the apparent acclimation of mesophyll photosynthetic processes. We conclude that it would be more effective to select for high photosynthetic capacity than for reduced stomatal sensitivity when breeding for increased drought resistance in black spruce seedlings.

The mechanism of 3-*O*-methyl-D-glucose transport through the plasmalemma has been investigated in protoplasts isolated from the mesophyll of *Pisum sativum* L. var. Dan.

Analysis of the fluxes after 50 minutes of uptake showed that the gradual decrease in slope of the net uptake curve with time was not due to any decline in uptake capacity; it represented the approach to flux equilibrium of a small compartment of the protoplast, probably the cytoplasm.

The energy of activation for initial flux into this

compartment was 20 kilocalories per mole between 17 and 27 C. Very high discrimination was shown with regard to sugar isomers. Light strongly promoted flux (by a factor of 2.5 in the case of methyl glucose). Initial flux showed sharply contrasting inhibitor sensitivity in the light and the dark. Light uptake was sensitive to the proton conductor carbonyl cyanide *m*-chlorophenylhydrazone (CCCP), but stable for at least the first 10 minutes to the ATPase inhibitors quercetin, rutin, and diethylstilbestrol, as well as to arsenate. Dark uptake, on the other hand, was stable to CCCP but was immediately depressed by quercetin, rutin, diethylstilbestrol, and arsenate.

Protoplasts which received a light pre-treatment before incubation in the dark took up methyl glucose at the accelerated light rate for the first 7 minutes. Moreover, the light pre-treatment sensitised subsequent initial dark uptake to CCCP, and conferred on it the stability to ATPase inhibitors and arsenate characteristic of light uptake. After about 7 minutes the characteristic inhibitor responses of dark uptake were resumed.

It is proposed that more than one mode of energy-coupling for sugar transport may operate in these protoplasts.

Possible Recycling of Photorespiratory CO_2

The system of protoplasts offers a good model to study photosynthesis. Mesophyll protoplasts of pea required only 1 mM bicarbonate for maximal photosynthesis, unlike chloroplasts, which required up to 10 mM bicarbonate. Such a markedly low requirement of bicarbonate for photosynthesis in mesophyll protoplasts was surprising, but it could be due to an internal carbon source and/or a CO_2 concentrating mechanism in mesophyll protoplasts. Ethoxyzolamide (EZA), inhibitor of cytosolic carbonic anhydrase (CA) suppressed photosynthesis at low bicarbonate (0.1 mM) but had no significant effect at high bicarbonate (10 mM). Acetazolamide (AZA), an inhibitor of plasma membrane CA, did not exert as much dramatic effect as EZA. Three photorespiratory inhibitors, aminoacetonitrile (AAN) or glycine hydroxamate (GHA) or aminooxyacetate (AOA) inhibited markedly photosynthesis at low bicarbonate, but not at high bicarbonate. Inhibitors of glycolysis or TCA cycle (NaF, sodium malonate) or PEP carboxylase (DCDP) had no significant effect on photosynthesis.

The bicarbonate requirement of protoplast photosynthesis and its sensitivity of photosynthesis to EZA (carbonic anhydrase inhibitor) were much higher at low oxygen (65 nmol ml^{-1}) than that at normal oxygen (212 nmol ml^{-1}). In contrast, the inhibitory effect of photorespiratory inhibitors on photosynthesis was similar in both normal and low oxygen medium. The marked elevation of glycine/serine ratio at low O_2 or in presence of GHA confirmed the suppression of photorespiratory decarboxylation by GHA. We suggest that photorespiration is a significant source of CO_2 for photosynthesis in mesophyll protoplasts at limiting CO_2 and at normal levels of oxygen.

Molecular Mechanism of Chloroplast Development Regulated by Plant Hormones

The chloroplast is a photosynthetic organelle in plant cells. The development and function of the chloroplast are regulated by plant hormones, particularly cytokinin as a positive regulator and brassinosteroid as a negative regulator. We would like to clarify the molecular mechanism of chloroplast development regulated by the two phytohormones, and attempts to clarify the mechanism using the approaches in molecular biology and molecular genetics are in progress.

Chloroplast is a photosynthesis apparatus that develops from proplastid. The chloroplast is a unique organelle in plant cells and plays important roles in primary metabolism, such as photosynthesis, nitrate assimilation and so on. The products of such metabolism, i.e., oxygen, carbohydrates and amino acids, are useful not only for the plant itself but also for most living things on the earth as basic requirements for aspiration and primary reaction of food chains. A chloroplast is one of differentiated type of the 'plastid'.

The origin of plastid is called as a 'proplastid', which possesses few characteristic structures and exists in seedlings in the developing stage and surface-layer cells of the apical meristem. In the photosynthetic mesophyll cells of leaves in the light, the proplastid differentiates into a 'chloroplast', which possesses a highly structured thylakoid membrane as a photosynthesis machinery. In contrast, the plastid of leaf mesophyll cells in the dark differentiates into an 'etioplast', which is characterised by paracrystalline tubules, and the plastid in root cells differentiates into an 'amyloplast', which is a starch-accumulating plastid. In a plant grown in the dark and then exposed to light, the etioplast develops into a chloroplast. The chloroplast also develops into a chromoplast in older staged leaves and fruits as pigment and secondary metabolite-accumulating plastids.

Chloroplasts have their own independent and unique genome of 150 kbp that encodes more than 100 genes. Any of them are highly involved in photosynthesis and they are regulated by transcriptional and other later step mechanisms.

Recently, RNA polymerase encoded on the chloroplast genome, the subunit sigma-factor encoded on the nuclear genome, and RNA polymerase encoded on the nuclear genome and transported into chloroplasts have been identified and their functional analyses are in progress. Then, we have identified and characterised two proteins that can bind to chloroplast DNA, and named them CND41, a 41-kDa DNA-binding protein purified from chloroplast nucleoids of photomixotrophically cultured tobacco cells, and PTF1, a transacting factor of the psbD enhancer domain from Arabidopsis.

The functional analysis had also revealed many aspects of the mechanism of chloroplast gene transcription. The assembly of the photosynthetic machinery in chloroplasts is highly coordinated by the proteins encoded on the plastid genome, and proteins encoded on nuclear genome and later imported into chloroplasts.

The combined expressions of both genomes for chloroplast were regulated by a number of factors, such as many environmental stimulations in depending on developmental mechanism during plant growing, and plant hormones as messengers in the plant of these conditions. We would like to clarify the molecular mechanism of plant hormone regulation related to the chloroplast development.

Cytokinin as a positive regulator of chloroplast development. Cytokinin is an adenine-derived plant growth regulator with important biological activities, including the promotion of cell division and chloroplast development. When dicot plants were grown in a medium containing cytokinin, the plants had short hypocotyls and highly greened leaves that correspond to the 'accelerated photomorphogenesis' phenotype.

The fundamental nature of these cytokinin effects suggests that changes in gene expression may be required in mediating the response of chloroplast development to cytokinin. Although most photosynthetic genes were induced by cytokinin, key genes.

Recently, we have determined that cultured green tobacco cells, *Nicotiana tabacum* cv. Samsun NN, are suitable for studying cytokinin signal transduction because cell growth and chloroplast development were up-regulated by a medium supplemented with cytokinin. Based on the developed fluorescent differential display using cultured green tobacco cells, the expression of three cytokinin-inducible genes (cig) during the earliest stages of chloroplast development were identified.

Functional analyses of the correlation between these cigs and chloroplast development are in progress. Other cytokinin-inducible genes will be able to be identified by another new techniques, i.e., use of micro arrays, in the near future.

Brassinosteroid as a negative regulator of chloroplast development: Brassinosteroid is a sterol-derived plant growth regulator with important biological activities, including the promotion of cell elongation and negative regulation of chloroplast development. Deficiency in brassinosteroid causes dwarf with highly greened leaves.

Recently, Asami and colleagues synthesised brassinosteroid biosynthesis inhibitor, brz 14. The brz-treated plants exhibited the same dwarf phenotype similar to DET2 and BRI1 mutants, which corresponds to the 'accelerated photomorphogenesis' phenotype. The brz-treated plants also had enhanced chloroplast development compared to the non-treated plants that could be results of the chloroplastic gene expression that encoded on nuclear and chloroplast genomes.

In order to analyse the mechanism of brassinosteroid signal transduction in chloroplasts and plant development, we screened for mutants that showed resistance to these brz effects. Screening of 200000 Arabidopsis seeds subjected to EMS and fast neutron mutagenesis revealed that some mutants were significantly taller than the wild type when grown in the dark with brz. At least three mutants grown in the dark with brz had hypocotyls as long as that of the wild types grown in the dark without brz treatment. These were named bil (brz-insensitive-long hypocotyl). Another three mutants had short hypocotyls but exhibited no cotyledon opening when grown in the dark with brz; these were named bih (brz-insensitive-hooked hypocotyl). In some of the mutants, pale greened leaves that might relate to chloroplast development were identified.

Chloroplasts and Photosynthesis

Chloroplasts are organelles found in plant cells and eukaryotic algae that conduct photosynthesis. Chloroplasts absorb sunlight and use it in conjunction with water and carbon dioxide to produce sugars. Chloroplasts capture light energy from the sun to conserve free energy in the form of ATP and reduce NADP to NADPH through a complex set of processes called photosynthesis. It is derived from the Greek words chloros which means green and plast which means form or entity. Chloroplasts are members of a class of organelles known as plastids.

Evolutionary Origin

Chloroplasts are one of the many unique organelles in the cell. They are generally considered to have originated as endosymbiotic cyanobacteria (i.e. blue-green algae). This was first suggested by Mereschkowsky in 1905. In this respect they are similar to mitochondria but are found only in plants and protista. The chloroplast is surrounded by a double-layered composite membrane with an intermembrane space; it has its own DNA and is involved in energy metabolism. Further, it has reticulations, or many infoldings, filling the inner spaces.

In green plants, chloroplasts are surrounded by two lipid-bilayer membranes. The inner membrane is now believed to correspond to the outer membrane of the ancestral cyanobacterium. The chloroplast genome is considerably reduced compared to that of free-living cyanobacteria, but the parts that are still present show clear similarities. Plastids may contain 60-100 genes whereas cyanobacteria often contain more than 1500 genes. Many of the missing genes are encoded in the nuclear genome of the host. The transfer of nuclear information has been estimated in tobacco plants at one gene for every 16000 pollen grains.

In some algae (such as the heterokonts and other protists such as Euglenozoa and Cercozoa), chloroplasts seem to have evolved through a secondary event of endosymbiosis, in which a eukaryotic cell engulfed a second eukaryotic cell containing chloroplasts, forming chloroplasts with three or four membrane layers. In some cases, such secondary endosymbionts may have themselves been engulfed by still other eukaryotes, thus forming tertiary endosymbionts.

Structure

Chloroplasts are observable morphologically as flat discs usually 2 to 10 micrometre in diameter and 1 micrometre thick. The chloroplast is contained by an envelope that consists of an inner and an outer phospholipid membrane. Between these two layers is the intermembrane space.

The material within the chloroplast is called the stroma, corresponding to the cytosol of the original bacterium, and contains one or more molecules of small circular DNA. It also contains ribosomes, although most of its proteins are encoded by genes contained in the cell nucleus, with the protein products transported to the chloroplast.

Within the stroma are stacks of thylakoids, the sub-organelles which are the site of photosynthesis. The thylakoids are arranged in stacks called grana (singular: granum). A thylakoid has a flattened disk shape. Inside it is an empty area called the thylakoid space or lumen. Photosynthesis takes place on the thylakoid membrane; as in mitochondrial oxidative phosphorylation, it involves the coupling of cross-membrane fluxes with biosynthesis via the dissipation of a proton electrochemical gradient.

Embedded in the thylakoid membrane is the antenna complex, which consists of proteins, and light-absorbing pigments, including chlorophyll and carotenoids. This complex both increases the surface area for light capture, and allows capture of photons with a wider range of wavelengths. The energy of the incident photons is absorbed by the pigments and funnelled to the reaction centre of this complex through resonance energy transfer. Two chlorophyll molecules are then ionised, producing an excited electron which then passes onto the photochemical reaction centre.

Plastids in Plants

Plastids are responsible for photosynthesis, storage of products like starch and for the synthesis of many classes of molecules such as fatty acids and terpenes which are needed as cellular building blocks and for the function of the plant. Depending on their morphology and function, plastids have the ability to differentiate, or redifferentiate, between these and other forms. All plastids are derived from proplastids (formerly "eoplasts", *eo-*: dawn, early), which are present in the meristematic regions of the plant. Proplastids and young chloroplasts commonly divide, but more mature chloroplasts also have this capacity. In plants, plastids may differentiate into several forms, depending upon which function they need to play in the cell. Undifferentiated plastids (*proplastids*) may develop into any of the following plastids:

- Chloroplasts: for photosynthesis; *see also etioplasts, the predecessors of chloroplasts*

- Chromoplasts: for pigment synthesis and storage
- Leucoplasts: for monoterpene synthesis; *leucoplasts sometimes differentiate into more specialised plastids:*

  — Amyloplasts: for starch storage
- Statoliths: for detecting gravity

  — Elaioplasts: for storing fat

  — Proteinoplasts: for storing and modifying protein

Each plastid creates multiple copies of the rectangular 75-250 kilo bases plastid genome. The number of genome copies per plastid is flexible, ranging from more than 1000 in rapidly dividing cells, which generally contain few plastids, to 100 or fewer in mature cells, where plastid divisions has given rise to a large number of plastids. The plastid genome contains about 100 genes encoding ribosomal and transfer ribonucleic acids (rRNAs and tRNAs) as well as proteins involved in photosynthesis and plastid gene transcription and translation. However, these proteins only represent a small fraction of the total protein setup necessary to build and maintain the structure and function of a particular type of plastid. Nuclear genes encode the vast majority of plastid proteins, and the expression of plastid genes and nuclear genes is tightly co-regulated to allow proper development of plastids in relation to cell differentiation.

Plastid DNA exists as large protein-DNA complexes associated with the inner envelope membrane and called 'plastid nucleoids'. Each nucleoid particle may contain more than 10 copies of the plastid DNA. The proplastid contains a single nucleoid located in the centre of the plastid. The developing plastid has many nucleoids, localised at the periphery of the plastid, bound to the inner envelope membrane. During the development of proplastids to chloroplasts, and when plastids convert from one type to another, nucleoids change in morphology, size and location within the organelle. The remodelling of nucleoids is believed to occur by modifications to the composition and abundance of nucleoid proteins.

In plant cells long thin protuberances called stromules sometimes form and extend from the main plastid body into the cytosol and interconnect several plastids. Proteins, and presumably smaller molecules, can move within stromules. Most cultured cells that are relatively large compared to other plant cells have very long and abundant stromules that extend to the cell periphery.

Plastids in Algae

In algae, the term leucoplast (leukoplast) is used for all unpigmented plastids. Their function differ from the leukoplasts in plants. Etioplast, amyloplast and chromoplast are plant-specific and do not occur in algae. Algal plastids may also differ from plant plastids in that they contain pyrenoids.

Inheritance of Plastids

Most plants inherit the plastids from only one parent. Angiosperms generally inherit plastids from the mother, while many gymnosperms inherit plastids from the father. Algae also inherit plastids from only one parent. The plastid DNA of the other parent is thus completely lost.

In normal intraspecific crossings (resulting in normal hybrids of one species), the inheritance of plastid DNA appears to be quite strictly 100 per cent uniparental. In interspecific hybridisations, however, the inheritance of plastids appears to be more erratic. Although plastids inherit mainly maternally in interspecific hybridisations, there are many reports of hybrids of flowering plants that contain plastids of the father.

Origin of Plastids

Plastids are thought to have originated from endosymbiotic cyanobacteria. Due to a split-up into three evolutionary lineages, the plastids are named differently: chloroplasts in green algae and plants, rhodoplasts in red algae and cyanelles in the glaucophytes. The plastids differ by their pigmentation, but also in ultrastructure. The chloroplasts, e.g. have lost all phycobilisomes, the light harvesting complexes found in cyanobacteria, red algae and glaucophytes, but — only in plants and in closely related green algae - contain stroma and grana thylakoids. The glaucocystophycean plastid — in contrast to the chloroplasts and the rhodoplasts — is still surrounded by a remains of the cyanobacterial cell wall. All these primary plastids are surrounded by two membranes.

Complex plastids start by secondary endosymbiosis, when a eukaryote engulfs a red or green alga and retains the algal plastid, which is typically surrounded by more than two membranes, and reduced in its metabolic and photosynthetic capacity. Algae with complex plastids derived by secondary endosymbiosis of a red alga include the heterokonts, haptophytes, cryptomonads, and most dinoflagellates (rhodoplasts). Those that endosymbiosed a green alga include the euglenids and chlorarachniophytes (chloroplasts). The Apicomplexa, a phylum of obligate parasitic protozoa including the causative agents of malaria (*Plasmodium* spp.), toxoplasmosis (*Toxoplasma gondii*), and many other human or animal diseases also harbour a complex plastid (although this organelle has been lost in some apicomplexans, such as *Cryptosporidium parvum*, which causes cryptosporidiosis). The 'apicoplast' is no longer capable of photosynthesis, but is an essential organelle, and a promising target for anti-parasitic drug development.

Some dinoflagellates take up algae as food and keep the plastid of the digested alga to profit from the photosynthesis; after a while the plastids are also digested. These captured plastids are known as kleptoplastids.

Chloroplasts are dynamic organelles within the plant cell. Movement of entire chloroplasts to maximise light absorption was observed long ago by light microscopy. Chloroplasts are surrounded by an envelope membrane, which sometimes appears to circulate around the main body of the organelle. Long thin protuberances sometimes form and extend from the main plastid body into the cytosol, occasionally touching and fusing with projections extending from other chloroplasts

(Kohler *et al.*, 1997b; Movie 1). For many years these tubular structures, which are now termed stromules, were neglected as a feature of chloroplasts because they are not present on all chloroplasts, are often not preserved in thin sections prepared for electron microscopy, and are difficult to observe by light microscopy. While occasional reports of envelope movement (described as a "mobile jacket") and chloroplast extensions appear sporadically in the literature of the 20th century (Gray *et al.*, 2001), the most convincing documentation of their existence was provided by S. Wildman and colleagues in the 1960 quotes (Wildman *et al.*, 1962).

After preparation of thin hand-sections of leaf cells, motile chloroplast protuberances were clearly documented by filming living cells in the phase microscope (Hongladarom *et al.*, 1964). Since then, similar structures have been detected by light microscopy of the single-cell eukaryotes Euglena gracilis (Ehara *et al.*, 1990) and Acetabularia (Menzel, 1994).

Not until plastids could be labelled in vivo with Green Fluorescent Protein (GFP) could stromules be readily observed on both chloroplasts and non-green plastids. The cloning of the gene for GFP, a protein from the jellyfish Aequorea victoria that fluoresces with no exogenous co-factor, has revolutionised studies of intracellular organelle and protein movement. The GFP coding sequence may be attached to another protein coding region and the resultant chimeric gene can be introduced into a genome to specifically label a protein or subcellular compartment (Hanson and Kohler, 2001).

The first experiment to label chloroplasts in vivo with GFP utilised a nuclear transgene carrying a chloroplast transit sequence fused to the GFP sequence. When the chimeric protein was synthesised in the transgenic plant cell, the chloroplast transit peptide directed the GFP to the chloroplast. It was possible to verify that the GFP was present in the chloroplast by first illuminating a plant cell with blue light that causes GFP to fluoresce green.

Filters allowed visualisation of the green light that was

emitted. The same cell was then illuminated again, and by using a different set of specific filters, only the red emitted light was collected and displayed. Merging the two images reveals that chlorophyll and GFP are both present in the main plastid body. However, only GFP, not chlorophyll, was detected in projections extending from the plastid body. These projections are not observed on chloroplasts of most leaf cells.

Nevertheless, if the GFP-encoding gene is expressed at sufficiently high levels, visual searching through a transgenic leaf will usually reveal at least a few cells containing chloroplasts with motile stromules. The absence of chlorophyll indicates that few, if any, thylakoid membranes enter the protuberance, hence the term "stroma-filled tubule" or stromule.

Before the era of GFP technology, non-green plastids were difficult to observe in vivo by light microscopy because they lack the green chlorophyll label. Non-green plastids may accumulate pigments (chromoplasts) or starch (amyloplasts), depending on the tissue where they are located. Electron micrographs of plastids in non-green tissues have sometimes revealed organelles whose appearance differs from the typical ovoid chloroplasts.

Fluorescence microscopy of a variety of organs of GFP-labelled transgenic plants makes it evident that non-green plastids are often irregular in shape and are much more likely to carry stromules than are chloroplasts. In some tissues, virtually every plastid is comprised partly of long thin tubular structures. Most cells in the tobacco petal, root, and hypocotyl contain plastids that exhibit stromules (Kohler and Hanson, 2000). Most cultured cells that are relatively large compared to other plant cells have very long and abundant stromules that extend to the cell periphery.

Proteins, and presumably smaller molecules, are known to move within stromules. A Fluorescence Loss After Photobleaching (FRAP) experiment proved that GFP could move from one plastid to another. In the FRAP experiment, the fluorescence of GFP in one of two interconnected plastids was destroyed by bleaching the GFP by a quick pulse of high-level light irradiation. The plastid that lost fluorescence by photobleaching became fluorescent again quickly when GFP flowed from the unirradiated plastid through the stromule to the bleached plastid (Kohler *et al.*, 1997a).

A technique termed Fluorescence Correlation Spectroscopy (FCS) could be used to measure the rate of movement of GFP molecules in stromules. FCS experiments revealed that GFP moves both actively and by diffusion through stromules, and that the stroma is very viscous. Both single molecules and batches of multiple GFP molecules, perhaps located within vesicles, could be detected (Kohler *et al.*, 2000). The brightly fluorescence points visible on the stromule in Movie 1 may represent such GFP "batches."

Now that the widespread presence of stromules has been documented in many types of plant tissues and species, we face the challenge of understanding their function. Stromules could possibly play a role in plastid movement, especially in extremely large cells such as those of Acetabularia. Stromules also increase the surface area of the plastid envelope, which is actively engaged in import and export of molecules to and from the cytoplasm.

The presence of stromules increases the surface area and extends the reach of an individual plastid to a part of the cell away from the main plastid body. Stromules potentially could act as conduits of materials from one part of the cell to a plastid body located elsewhere. Stromules may also be involved in intracellular coordination because they are often located close to other organelles such as mitochondria and nuclei, and sometimes connect different plastids to one another.

The Photophosphorylation

ATP is the energy currency of life. Production of glucose (food) from CO_2 requires both NADPH and ATP.

After indications in several laboratories that algal cells produce ATP in light, Arnon *et al.* (1954) and Frenkel (1954) discovered photophosphorylation in chloroplasts of plants, and chromatophores of anoxygenic photosynthetic bacteria, respectively. This was a major breakthrough and it took many years to recognise that *only* in some anoxygenic photosynthetic bacteria, almost all light energy is first converted into ATP energy and then this energy is used for the reversed electron flow to produce the reducing power in the form of NADH. In oxygenic organisms, however, this is not the case as correctly expressed by Rabinowitch (1956).

As we know today, ATP synthesis follows electron transport steps that first store energy temporarily by creating a proton motive force; this energy would be otherwise lost if not used for ATP formation. P. Mitchell (1961), who received the Nobel prize in Chemistry in 1978, provided the theory that a proton motive force (that is a sum of a pH gradient and a membrane potential) is the energy source of ATP synthesis. Membranes are normally impermeable to protons; and, protons are transferred from one side of the membrane to the other by virtue of the alternate electron and hydrogen atom transfers due to specific location of the electron and hydrogen atom carriers and the energetics of the light-induced electron transfers in photosynthesis.

An early breakthrough was the independent experiment of Shen & Shen (1962) and of Hind & Jagendorf (1963) showing that light forms some entity (X_E) that can be used later in darkness to produce ATP. Hind & Jagendorf showed that X_E is a pH gradient that drives ATP synthesis. Jagendorf & Uribe (1966) showed that ATP can be synthesised from pH gradient created by acid-base transition in total darkness. These experiments, along with those of the others, provided major evidence for the chemiosmotic hypothesis of Mitchell. In photosystem II, water oxidation complex that liberates protons is on the inner side (the lumen) of the thylakoid membranes. Further, plastoquinone is reduced to plastoquinol on the outer side (stroma side) of the thylakoid membrane; the plastoquinol (a hydrogen atom carrier) moves towards Rieske Iron sulphur and cytochrome f (both are electron, not hydrogen atom carriers) that are located towards the lumen side. Here, plastoquinol delivers its electrons to the Rieske Iron sulphur centre and the cytochrome f, leaving the protons to be released into the lumen.

This arrangement, thus, allows natural proton translocation from the stroma to the lumen side as the electron transport takes place in PSII. This adds to the pH gradient to be used for ATP synthesis. The equivalence of pH gradient and membrane potential in synthesising ATP was shown by the elegant experiments of Hangarter & Good (1982) when they varied one keeping the other constant and showing that it was the sum of the two that was important for initiating ATP synthesis.

The understanding of the mechanism by which ATP synthesis takes place at the ATP synthase using the proton motive force has been influenced by the following three breakthroughs.

(1) The binding change hypothesis of P. Boyer and co-workers that suggests that at one time

(a) ADP and Pi (inorganic phosphate) bind weakly at one site of the alpha-beta pair of the F1 part of the ATP synthase enzyme;

(b) Bound ATP is formed from bound ADP and Pi, without the use of energy at a second alpha-beta pair; and

(c) The pH gradient energy is converted into rotational energy, mainly of the gamma subunit that extends upto the third alpha-beta pair, that is used to flip off the ATP free from the third alpha-beta binding site. These three sites alternate in time.

(2) The rotation feature of the Boyer model has been elegantly demonstrated directly by fluorescence microscopy and by photoselection and other experiments of beef-heart mitochondrial F1 showed that the gamma subunit indeed looks through the alpha-beta pairs; and in agreement with the Boyer model, the structure shows one alpha-beta pair site empty (as if ATP was released); another with bound ADP and Pi; and the third with an equivalent of bound ATP.

The mechanism of photosynthesis has been probed by several means. Its face has changed by experiments of many investigators. Using the repetitive flash method and Warburg's manometry, they led to the concept of "Photosynthetic Unit" where excitation energy, absorbed by hundreds of antenna pigment molecules, is transferred to the reaction centre chlorophyll molecules, the "photoenzyme" or the "energy trap" for chemistry. Next, was the discovery of the "Red Drop" (Emerson & Lewis, 1943) in the quantum yield action spectrum of oxygen evolution in the green alga *Chlorella pyrenoidosa*, and the enhancement effect of certain wavelengths of light on the yield of oxygen evolution in the "red drop" region (Emerson *et al.*, 1957), discovered by the use of an excellent monochromator and state-of-the art manometry; it lead to the concept of two-light reaction and two-pigment system mechanism of electron transport. It became well-known due to the working hypothesis of Hill & Bendall (1960) and became a scientific fact by the experiments of Duysens *et al.* (1961) on the antagonistic effect of light absorbed in pigment system I and II on the redox state of cytochrome *f*. The important concept that there is an "oxygen clock", where four positive charges must accumulate before water can be oxidised to oxygen was enshrined by:

(a) The experiments of Joliot *et al.* (1969) on the periodicity of four in the plots of the amount of oxygen evolved per flash as a function of flash number, and

(b) The so-called "S-states" model of charge accumulation enunciated by Kok *et al.* (1970).

Such a periodicity of four, reflecting indirectly the S-states, has also been observed in chlorophyll *a* fluorescence (Delosme, 1971; Joliot & Joliot, 1971), and recently, Shinkarev *et al.* (1997) have even managed to obtain the kinetics of the last step of oxygen evolution from analysis of the data on the decay of chlorophyll *a* fluorescence in single flashes of light. Thermoluminescence, discovered by W. Arnold & H. Sherwood, that has been exploited both by P.V. Sane (Sane & Rutherford, 1986) as well as (DeVault *et al.*, 1983; Vass & Govindjee, 1996) has uniquely probed the characteristics of the S-states of oxygen evolution (Inoue & Shibata, 1978; Inoue, 1996). The most elegant probes for showing that Kok's S-states are Manganese at the low temperature EPR (Dismukes& Siderer, 1980); and the Extended X-ray Absorption Fine Structure (EXAFS) spectroscopy. Two elegant conceptual work on another area, that of phosphorylation:

(a) The chemiosmotic theory of P. Mitchell in which proton motive force across a membrane provide energy for ATP synthesis (Mitchell, 1961) and

(b) The elegant theory of Paul Boyer as to how this comes about: by conversion of electrochemical energy to conformational energy. Mimicking photosynthesis *in vitro* has been a dream of many, and the success of Steinberg-Yfrach *et al.* (1998) in making "lots of ATP" in artificial liposome membranes, energised by a synthetic system (carotene-porphyrin-quinone, is highly commendable.

Photosynthesis and 'Z' Scheme

Reading the Z-scheme: Whenever molecules gain or lose electrons energy is involved. The Z-scheme is an energy diagram for electron transfer in the "light reactions" of plant photosynthesis. It applies equally well to photosynthesis by algae and cyanobacteria. The vertical energy scale shows each molecule's ability to transfer an electron to (i.e., to reduce) the next one from left to right. The ones at the top transfer electrons easily to the ones below them as it is a "downhill" reaction, energy-wise. However, for electron transfer from those at the bottom to those above them it is an "uphill" reaction and requires input of outside energy.

The Z scheme shows the pathway of electron transfer from water to $NADP^+$. Using this pathway, plants transform light energy into "electrical" energy (electron flow) and hence into chemical energy as reduced NADPH and ATP. Later in the "dark reactions" of photosynthesis, that chemical energy is further transformed into the chemical bonds in sugar molecules. In the complete process, carbon dioxide is joined into the sugar molecules and oxygen is released. Although it is not a structural diagram, the Z-scheme does give the sequence of electron flow (oxidation and reduction) with an energy perspective. The source of electrons is water (H_2O). The large vertical red arrows represent excitation of reaction centre chlorophyll molecules (by light energy) and the black arrows represent electron flow, which is downhill energywise. This particular version of the "Z-scheme" was developed for the novel, THE MUSIC OF SUNLIGHT by *Sunlight Books* in collaboration with the authors of this article.

Mn is the manganese centre, a complex containing 4

manganese atoms, which splits two water molecules at a time into 4 protons ($4H^+$), 4 electrons ($4e^-$), and 2 oxygen atoms, as an oxygen molecule (O_2). Tyr is a special tyrosine molecule, also sometimes referred to as Yz or simply as Z, which shuttles electrons to the "reaction centre" of photosystem II (PS II). Chl P680 is the reaction centre pair of chlorophyll *a* molecules of PS II. Excited Chl P680* has reached this state by absorbing a photon of light energy. Pheo is a pheophytin molecule, which is a chlorophyll with its central magnesium ion (Mg^{++}) having been replaced by two hydrogens.

It is the primary electron acceptor of PS II, whereas P680 is the primary electron donor. Q_A is a plastoquinone molecule, which is tightly-bound and immovable. It is known in some circles as the primary stable electron acceptor of PS II, and it accepts and transfers one electron at a time. Q_B is a loosely-bound plastoquinone molecule, which accepts two electrons and then takes on two protons, before it detaches and becomes mobile and called PQ. PQ is the detached plastoquinone molecule, which is mobile within the hydrophobic core of the thylakoid membrane. FeS is the Rieske iron-sulphur protein. Cyt f is cytochrome f. Cyt b_{6L} and Cyt b_{6H} are two cytochrome, b_6 molecules (of lower and higher energy). PC is plastocyanin, a highly mobile copper protein. Chl P700 and Excited Chl P700* arc respectively the ground energy state and the excited energy state of the chlorophyll molecule of the "reaction centre" of photosystem I. A0 is a special chlorophyll *a* molecule that is the primary electron acceptor of PSI, whereas P700 is the primary electron donor of PSI. A_1 is a phylloquinone (vitamin K) molecule. F_X F_A and F_B are three separate immobile iron-sulphur protein centres. FD is ferredoxin, a somewhat mobile iron-sulphur protein. FNR is the enzyme ferredoxin-NADP oxidoreductase, which contains the active group, called FAD (flavin adenine dinucleotide). $NADP^+$ is the oxidised form of nicotinamide adenine dinucleotide phosphate. NADPH is its reduced form

In plants (as well as algae and cyanobacteria), photosynthesis has two major phases:

1. The Light Reactions comprise the light-dependent phase, which produces the reducing power ("reducing", as in "oxidation and reduction"), ATP (adenosine triphosphate – the energy currency of life), and oxygen (O_2). This all takes place in and around the thylakoid membranes.
2. The Dark Reactions are not directly dependent on the presence of light. Here the reducing power of NADPH and the energy of ATP (both generated by the light reactions) are used to convert, or "fix" CO_2 into sugars. These reactions occur in the stroma matrix and are called the Calvin-Benson cycle or C_3 cycle.

The Z-scheme represents the steps in the light reactions, showing the pathway of electron transport from water to $NADP^+$ (nicotinamide adenine dinucleotide phosphate). This leads to the release of oxygen, the "reduction" of $NADP^+$ to NADPH (by adding two electrons and one proton), and the building-up of a high concentration of hydrogen ions inside the thylakoid lumen (needed for ATP production).

Why is it called Z-Scheme?

It is simply because the diagram, when first drawn, was in the form of the letter "Z". Currently, it is being drawn to emphasise the energy levels of the components. Thus, it has been turned 90 degrees counter-clockwise.

Operation of Z-Scheme

Excitation of Reaction Centres: Photosynthesis starts with the simultaneous excitation of pairs of special reaction centre chlorophyll(a) molecules, labelled as P680 (in photosystem II, or PS II) and P700 (in photosystem I, or PSI). The excitation energy comes either from directly absorbed light or (most often) by energy transfer from adjacent pigment molecules in protein complexes called antennas. These "antenna" pigment molecules (chlorophylls and carotenoids) absorb light energy and then transmit it by inductive resonance from one molecule to the next, finally to the reaction centre. Excitation is over within a few femtoseconds (10^{-15} s). [A second has as many femtoseconds in it as 31 million years has seconds.]

The First Chemical Step: Normally, one describes the Z-diagram from left to right as if water delivers electrons first. This is not correct, but it is easy to describe the process that way. Thus, most books just do that. We shall, however, describe the steps as they appear in an actual approximate time sequence. The first chemical step happens within only a few picoseconds (10^{-12} s) when excited P680* loses an electron to Pheo, producing oxidised P680 ($P680^+$) and reduced Pheo ($Pheo^-$) in PS II, and excited P700* loses an electron to Ao, producing oxidised P700 ($P700^+$) and reduced Ao (Ao^-). This is the only step where light energy is converted to chemical energy, precisely oxidation-reduction energy. The rest of the steps are downhill energy-wise.

The Electron Transfer Steps: A molecule with a plus (+) charge has one less electron than its counterpart and is said to be the oxidised species, as it has lost one electron. The species that has an added electron is called the reduced species. *Reduction* means the addition of electrons or of hydrogen atoms {One hydrogen atom is a combination of one proton (H^+) and one electron (e^-)}, and *oxidation* means removal of either H or e^-. This oxidation and reduction process is what drives the activity in the Z-scheme sequence. Molecules higher on the diagram are able to reduce (pass an electron to) the next molecule lower down on the energy scale.

The recovery (reduction) of $P680^+$ to P680 and of $P700^+$ to P700 happens almost simultaneously. $P700^+$ recovers to P700 by receiving an electron that was passed down from reduced Pheo to Q_A (which is bound to the reaction centre protein complex), then to Q_B (another bound plastoquinone molecule). When Q_B has accepted two electrons from Q_A, it also takes on two protons from the stroma. Then it detaches from its protein binding site and diffuses through the hydrophobic core of the thylakoid membrane to the Cyt bf complex, where the electrons are passed on to an iron-sulphur protein (FeS, the Rieske protein) and then to a mobile copper-protein (PC, or plastocyanin) which finally carries a single electron to the oxidised P700+. Thus the electron is passed in ,"bucket brigade" manner through the "intersystem chain of electron (or H-atom) carriers"

There is a protein complex called the Cyt bf complex which contains FeS, Cytochrome f, and two cytochrome b_6 molecules. The "bottleneck", or the slowest step of the entire sequence, is the passage of the reduced PQ molecule to the Cyt bf complex and PQ's oxidation by FeS. This takes several milliseconds (10^{-3} s). Cytochrome b_6 is active in the Q-cycle.

In PSI the electron on A_O^- is passed ultimately to $NADP^+$ via several other intermediates: A_1, a phylloquinone (vitamin K); F_X F_A and F_B which are immobile (bound) iron-sulphur proteins; ferredoxin, which is a somewhat mobile iron-sulphur protein molecule; and the enzyme ferredoxin-NADP reductase (FNR) which is actually an oxido-reductase and whose active group is FAD (short for flavin adenine dinucleotide).

The missing electron on $P680^+$ is recovered, ultimately, from water molecules on the left bottom via an amino acid tyrosine (a specific one in a protein of PS II, also referred to as Yz in the literature) and a tetra-nuclear manganese (Mn) cluster. These reactions also require a few milliseconds.

A total of 8 quanta (photons) of light (4 in PS II and 4 in PSI), are required to transfer 4 electrons from 2 molecules of water to 2 molecules of $NADP^+$. This produces 2 molecules of NADPH and 1 molecule of O_2. This is the oxygen that both plants and animals need for respiration and life.

Proton Gradient and ATP Synthesis

The light reactions provide not only the reducing power in NADPH but also the energy of ATP, both essential for producing sugars from CO_2. ATP is produced through an enzyme called ATP synthase, using ADP (adenosine diphosphate), inorganic phosphate (P_i) and energy from a proton motive force (pmf) across the thylakoid membrane. This pmf is composed of two components: an electrical potential across the thylakoid membrane and a proton gradient across the thylakoid membrane. The proton gradient comes from the storing up of protons (hydrogen ions) inside the lumen, giving a pH of 6 inside the lumen and pH of 8 outside, in the stroma. Then, basically, protons escaping out from the thylakoid lumen through a central core of the enzyme ATP synthase (embedded in the membrane) cause conformational (rotational) changes in the enzyme, which catalyses the phosphorylation of ADP and the release of ATP on the stromal side.

Protons (hydrogen ions) are concentrated into the lumen in several ways:

1. Oxidation of water not only releases O_2 and "sends" electrons to P680, but it also releases protons (H^+) into the lumen.
2. When Q_B is reduced in PS II, it not only receives two electrons from Q_A but it also picks up two protons from the stroma matrix and becomes PQH_2. It is able to "carry" both electrons and protons (hydrogens). It is a H-atom carrier. At the Cyt bf complex it is then oxidised, but FeS and Cyt b_6 can only accept electrons (not protons). So the two protons are released into the lumen.
3. The Q-cycle of the Cyt bf complex is great because it provides extra protons into the lumen. Here two electrons travel through the two haemes of cytochrome b_6 and then reduce PQ on the stroma side of the membrane. The reduced PQ takes on two protons from the stroma, becoming PQH_2, which migrates to the lumen side of the Cyt bf complex where it is again oxidised, releasing two more protons into the lumen. Thus the Q-cycle allows formation of more ATP.
4. When $NADP^+$ is reduced by two electrons, it also picks up one proton, in effect removing it from the stroma and further increasing the gradient across the membrane.

Historical Origin of Z-Scheme

The Z-scheme owes its origin to several investigators. First it was Robert Emerson and his co-workers, including the authors (1956-1964) at the University of Illinois (at Urbana-Champaign) who discovered and embellished the "Enhancement effect" in oxygen evolution, which occurred when light absorbed in one photosystem (currently called PSI) was added to light absorbed in another photosystem (currently called PS II). Experiments using chloroplasts, and those using a mass spectrometer, absorption spectrometer, a fluorometer and electron spin resonance spectrometer were crucial to the establishment of the concepts. These suggested the existence of "two light reactions and two photosystems".

It was Bessel Kok and co-workers (1959-1960) at Baltimore, Maryland, and Lou NM Duysens, J. Amesz and co-workers (1959-1961) in Leiden, The Netherlands, who discovered the crucial antagonistic effect of light absorbed in PSI and PS II on the oxidation-reduction state of P700 and of cytochrome f (Cyt f, the electron carrier in the middle of the intersystem chain of intermediates), respectively. Duysens's experiments established the "series" nature of the present scheme. Light captured by PSI, in the "Z"-scheme, leads to oxidation of Cyt f (i.e., takes an electron away from it and places it on, say, $NADP^+$), whereas when light is captured by PS II, oxidised Cyt f is reduced by an electron coming from PS II. The theoretical concepts of Robin Hill and Fay Bendall (1960) (who essentially gave us the Z-scheme), the earlier schemes in books by Eugene Rabinowitch (1945-1956), and the detailed work of Horst Witt and co-workers (1961) in Berlin, Germany, played important and crucial roles in the origin of the "Z-scheme". The final evidence of its validity came from state-of-the-art detailed biophysics, biochemical, molecular biology, and genetic research in about 20 laboratories around the world. This is not to say that mutants and organisms may not be found in the future, that use alternate means to transfer electron from water to $NADPH^+$.

Photosystem

Photosystems (ancient Greek: phos = light and systema = assembly) are protein complexes involved in photosynthesis. They are found in the thylakoid membranes of plants, algae and cyanobacteria (in plants and algae these are located in the chloroplasts), or in the cytoplasmic membrane of photosynthetic bacteria. A photosystem (or Reaction Centre) is an enzyme which uses light to reduce molecules. This membrane protein complex is made of several subunits and contains numerous co-factors. In the photosynthetic membranes, reaction centres provide the driving force for the bioenergetic electron and proton transfer chain. When light is absorbed by a reaction centre (either directly or passed by neighbouring pigment-antennae), a series of oxido-reduction reactions is initiated, leading to the reduction of a terminal acceptor.

Two families of photosystems exist: type I reaction centres (like photosystem I (P700) in chloroplasts and in green-sulphur bacteria) and type II reaction centres (like photosystem II (P680) in chloroplasts and in non-sulphur purple bacteria). Each photosystem can be identified by the wavelength of light to which it is most reactive (700 and 680 nanometres, respectively for PSI and PS II in chloroplasts), and the type of terminal electron acceptor. Type I photosystems use ferredoxin-like iron-sulphur cluster proteins as terminal electron acceptors, while type II photosystems ultimately shuttle electrons to a quinone terminal electron acceptor. One has to note that both reaction centres types are present in chloroplasts and cyanobacteria, working together to form a unique photosynthetic chain able to extract electrons from water, evolving oxygen as a by-product.

Structure

A reaction centre comprises several (> 10) protein subunits, providing a scaffold for a series of co-factors. The latter can be pigments (like chlorophyll, pheophytin, carotenoids), quinones or iron-sulphur clusters. Because chlorophyll *a* can only absorb light of a narrow wavelength, it works with the antenna pigments to gain energy from a larger part of the spectrum. The pigments absorb light of various wavelengths and pass along their gained energy to the reaction centre chlorophyll. When the energy reaches the chlorophyll *a*, it releases two electrons into an electron transport chain.

Though chlorophyll *a* normally has an optimal absorption wavelength of 660 nanometres, it associates with different proteins in each type of photosystem to slightly shift its optimal wavelength, producing two distinct photosystem types. Other proteins serve to support the structure and electron pathways in the photosystem.

Relationship between Photosystems I and II

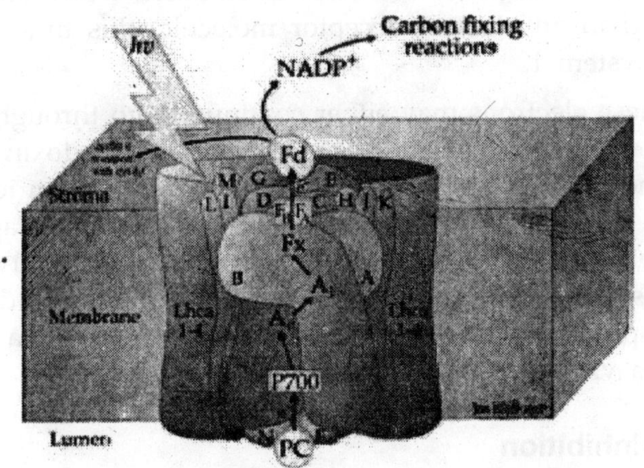

Schematic drawing of photosystem I from higher plants

Historically photosystem I was named one since it was discovered before photosystem II but this does not represent the order of the electron flow.

When photosystem II absorbs light, electrons in the reaction-centre chlorophyll are excited to a higher energy level and are trapped by the primary electron acceptors. To replenish the deficit of electrons, electrons are extracted from water (either through photolysis or enzymatic means) and supplied to the chlorophyll.

Photoexcited electrons travel though the cytochrome b6f complex to photosystem I via an electron transport chain set in the thylakoid membrane.

This energy fall is harnessed (the whole process termed chemiosmosis), to transport hydrogen (H+) through the membrane to provide a proton-motive force to generate ATP. If electrons only pass through once, the process is termed non-cyclic photophosphorylation.

When the electron reaches photosystem I, it fills the electron deficit of the reaction-centre chlorophyll of photosystem I. The deficit is due to photo-excitation of electrons which are again trapped in an electron acceptor molecule, this time that of photosystem I.

These electrons may either continue to go through cyclic electron transport around PS I, or pass via ferredoxin, to the enzyme NADP+ reductase. Electrons and hydrogen ions are added to NADP+ to form NADPH. This reducing agent is transported to the Calvin cycle to react with glycerate 3-phosphate, along with ATP to form glyceraldehyde 3-phosphate, the basic building block from which plants can make a variety of substances.

Photoinhibition

Photoinhibition is a reduction in a plant's (or other photosynthetic organism's) capacity for photosynthesis caused by exposure to strong light (above the saturation point). Photoinhibition is not caused by high light *per se*, but rather absorption of too much light energy compared with the photosynthetic capacity, i.e. any excess energy that the photosystem cannot handle is damaging. Too much light energy affects (photosystem II (P680)) more than photosystem I (PSI), and it has been hypothesised that the excess energy damages either the oxidising (donor) or reducing (acceptor) side of PS II, damaging the water oxidising complex on the donor side or blocking the flow of electrons to plastoquinone on the acceptor side (Hall & Rao 1999). Photoinhibition is often reversible, i.e. dynamic photoinhibition, and does in that case not inflict permanent damage to the photosystem. However, severe photoinhibition over a long time may cause highly reactive free oxygen radicals to form, which degrade photosynthetic components, i.e. chronic photoinhibition or photodamage. Particularly vulnerable is one of the main core proteins of photosystem II, protein D1, encoded by the gene psbA. Plants and algae have several mechanisms to protect against photoinhibition, e.g. through the xanthophyll cycle.

Plastoquinone

H_3C, H_3C, O, O, CH_3, $(CH_2-CH=C-CH_2)_9-H$

Plastoquinone (oxidised form)

Plastoquinone, (often abbreviated PQ), is a quinone molecule involved in the electron transport chain in the light-dependent reactions of photosynthesis. Plastoquinone accepts two atoms of hydrogen, becoming plastoquinol. It then transports those atoms to the cytochrome b6/f protein complex.

OH, H_3C, H_3C, OH, CH_3, $(CH_2-CH=C-CH_2)_9-H$

Plastoquinol (reduced form)

Semiquinone

A Ubisemiquinone is a free radical resulting from the removal of one hydrogen atom with its electron during the process of dehydrogenation of a hydroquinone to quinone or alternatively the addition of a single H atom to a quinone.

ATP Synthase

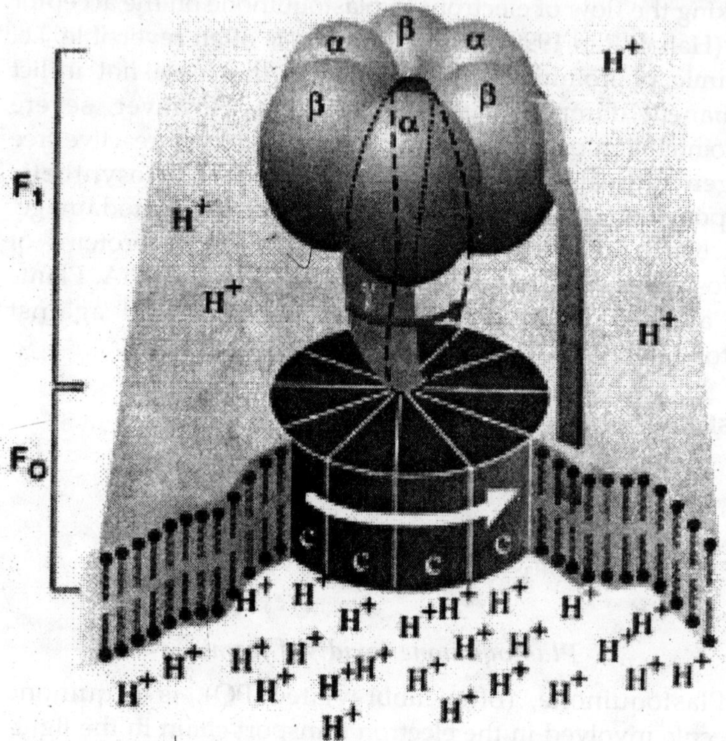

Figure: ATP synthase in E. coli

An ATP synthase is a general term for an enzyme that can synthesise adenosine triphosphate (ATP) from adenosine diphosphate (ADP) and inorganic phosphate by utilising some form of energy. The overall reaction sequence is:

$$ADP + P_i \longrightarrow ATP$$

These enzymes are of crucial importance in almost all organisms, because ATP is the common "energy currency" of cells.

In mitochondria, the F_1F_0 ATP synthase has a long history of scientific study. The F_1 portion of the ATP synthase is above the membrane, the F_0 portion is within the membrane. It's easy to visualise the F_0F_1 particle as resembling the fruiting body of a common mushroom, with the head being the F_1 particle, the stalk being the gamma subunit of F_1, and the base and "roots" being the F_0 particle embedded in the membrane. The F_1 particle was first isolated by Ephraim Racker in 1961. The nomenclature of the enzyme suffers from a long history.

The F_1 fraction derives it name from the term "Fraction 1" and F_0 (written as a subscript "O", not "zero") derives it name from being the oligomycine binding fraction. Taking as an example the nomenclature of subunits in the bovine enzyme, many subunits have Greek and Roman alphabet names (alpha, beta, gamma, delta, epsilon and a, b, c, d, e, f, g, h), while others have more complex names such as F_6 (from "Fraction 6"), OSCP (the oligomycin sensitivity conferral protein), A6L (named for the gene that codes for it in the mitochondrial genome) and IF1 (inhibitory factor 1).

The F_1 particle is large and can be seen in the transmission electron microscope by negative staining (1962, Fernandez-Moran et al., Journal of Molecular Biology, Vol 22, p 63). These are particles of 9 nm diameter that pepper the inner mitochondrial membrane. They were originally called elementary particles and were thought to contain the entire respiratory apparatus of the mitochondrion, but through a long series of experiments, Ephraim Racker and his colleagues were able to show that this particle is correlated with ATPase activity in uncoupled mitochondria and with the ATPase activity in submitochondrial particles created by exposing mitochondria to ultrasound. This ATPase activity was further associated with the creation of ATP by yet another long series of experiments in many laboratories. The antibiotic oligomycin inhibits ATP synthase.

Binding Change Mechanism

In the 1960s through the 1970s, Paul Boyer developed his binding change, or flip-flop, mechanism, which postulated that ATP synthesis is coupled with a conformational change in the ATP synthase generated by rotation of the gamma subunit. The research group of John E. Walker, then at the MRC Laboratory of Molecular Biology in Cambridge but now at the MRC Dunn Human Nutrition Unit (also in Cambridge) crystallised the F_1 catalytic-domain of ATP synthase.

The structure, at the time the largest asymmetric protein structure known, indicated that Boyer's rotary-catalysis model was essentially correct. For elucidating this Boyer and Walker shared half of the 1997 Nobel Prize in Chemistry. Jens Christian Skou received the other half of the Chemistry prize that year "for the first discovery of an ion-transporting enzyme, Na+, K+ -ATPase"

The crystal structure of the F_1 showed alternating alpha and beta subunits (3 of each), arranged like segments of an orange around an asymmetrical gamma subunit. According to the current model of ATP synthesis (known as the alternating catalytic model), the proton-motive force across the inner mitochondrial membrane, generated by the electron transport chain, drives the passage of protons through the membrane via the F_O region of ATP synthase. A portion of the F_O (the ring of c-subunits) rotates as the protons pass through the membrane. The c-ring is tightly attached to the asymmetric central stalk (consisting primarily of the gamma subunit) which rotates within the $alpha_3beta_3$ of F_1 causing the 3 catalytic nucleotide binding sites to go through a series of conformational changes that leads to ATP synthesis. The major F_1 subunits are prevented from rotating in sympathy with the central stalk rotor by a peripheral stalk that joins the $alpha_3beta_3$ to the non-rotating portion of F_O. The structure of the intact ATP synthase is currently known at low-resolution from electron cryo-microscopy (cryo-EM) studies of the complex. The cryo-EM model of ATP synthase shows that the peripheral stalk is a flexible rope-like structure that wraps around the complex as it joins F_1 to F_O. Under the right conditions, the enzyme reaction can also be carried out in reverse, with ATP hydrolysis driving proton pumping across the membrane.

Physiological Role

Like other enzymes, the activity of F_1F_O ATP synthase is reversible. Large enough quantities of ATP cause it to create a transmembrane proton gradient, this is used by fermenting bacteria which do not have an electron transport chain, and hydrolyse ATP to make a proton gradient, which they use for flagella and transport of nutrients into the cell.

In respiring bacteria under physiological conditions, ATP synthase generally runs in the opposite direction, creating ATP while using the proton-motive force created by the electron transport chain as a source of energy. The overall process of creating energy in this fashion is termed oxidative phosphorylation. The same process takes place in mitochondria, where ATP synthase is located in the inner mitochondrial membrane (so that F_1-part sticks into mitochondrial matrix, where ATP synthesis takes place).

ATP Synthase in Different Organisms

Plant ATP Synthase: In plants ATP synthase is also present in chloroplasts (CF_1F_O-ATP synthase). The enzyme is integrated into thylakoid membrane; the CF_1-part sticks into stroma, where dark reactions of photosynthesis (also called the light-independent reactions or the Calvin cycle) and ATP synthesis take place. The overall structure and the catalytic mechanism of the chloroplast ATP synthase are almost the same as those of the mitochondrial enzyme. However, in chloroplasts the proton motive force is generated not by respiratory electron transport chain, but by primary photosynthetic proteins.

E. coli ATP synthase: *E. coli* ATP synthase is the simplest known form of ATP synthase, with 8 different subunit types.

Yeast ATP Synthase: Yeast ATP synthase is the most complex known and is made of 20 different types of subunits.

Evolution of ATP Synthase: The evolution of ATP synthase is thought to be an example of modular evolution, where two subunits with their own functions have become associated and gained new functionality. The F_O particle shows significant similarity to hexameric DNA helicases and the F_1 particle shows some similarity to H^+ powered flagellar motor complexes.

The $\alpha_3 \beta_3$ hexamer of the F_O particle shows significant structural similarity to hexameric DNA helicases; both form a ring with 3 fold rotational symmetry with a central pore. Both also have roles dependent on the relative rotation of a macromolecule within the pore; the DNA helicases use the helical shape of DNA to drive their motion along the DNA molecule and to detect supercoiling whilst the $\alpha_3 \beta_3$ hexamer uses the conformational changes due rotation of the γ subunit to drive an enzymatic reaction.

The H^+ motor of the F_1 particle shows great functional similarity to the H^+ motors seen in flagellar motors. Both feature a ring of many small alpha helical proteins which rotate relative to nearby stationary proteins using a H^+ potential gradient as an energy source. This is, however, a fairly tenuous link - the overall structure of flagellar motors is far more complex than the F_1 particle and the ring of rotating proteins is far larger, with around 30 compared to the 10, 11 or 14 known in the F_1 complex.

The modular evolution theory for the origin of ATP synthase suggests that two subunits with independent function, a DNA helicase with ATPase activity and a H^+ motor, were able to bind, and the rotation of the motor drive the ATPase activity of the helicase in reverse. This would then evolve to become more efficient, and eventually develop into the complex ATP synthases seen today. Alternatively the DNA helicase/ H^+ motor complex may have had H^+ pump activity, the ATPase activity of the helicase driving the H^+ motor in reverse. This could later evolve to carry out the reverse reaction and act as an ATP synthase.

Function

The ATP synthase enzymes have been remarkably conserved through evolution. The bacterial enzymes are essentially the same in structure and function as those from mitochondria of animals, plants and fungi, and the chloroplasts of plants. The early ancestory of the enzyme is seen in the fact that the Archaea have an enzyme which is clearly closely related, but has significant differences from the Eubacterial branch. The H^+-ATP-ase found in vacuoles of the eukaryote cell cytoplasm is similar to the archaeal enzyme, and is thought to reflect the origin from an archaeal ancestor.

In most systems, the ATP synthase sits in the membrane (the "coupling" membrane), and catalyses the synthesis of ATP from ADP and phosphate driven by a flux of protons across the membrane down the proton gradient generated by electron transfer. The flux goes from the protochemically positive (P) side (high proton electrochemical potential) to the protochemically negative (N) side. The reaction catalysed by ATP synthase is fully reversible, so ATP hydrolysis generates a proton gradient by a reversal of this flux. In some bacteria, the main function is to operate in the ATP hydrolysis direction, using ATP generated by fermentative metabolism to provide a proton gradient to drive substrate accumulation, and maintain ionic balance.

$$ADP + Pi + nH^+_P <=> ATP + nH^+_N$$

Because the structures seen in EM, the subunit composition, and the sequences of the subunits appeared to be so similar, it had been assumed that the mechanisms, and hence the stoichiometries, would be the same. In this context, the evidence suggesting that the stoichiometry of H^+/ATP (n above) varied depending on system was surprising. Values based on measure $ATP/2e^-$ ratios, and $H^+/2e^-$ ratios had suggested that n was 3 for mitochondria, and 4 for chloroplasts, but these values were based on the assumption of integer stoichiometries. Although all the F_1F_0-type ATP-synthases likely had a common origin, both the assumption that the stoichiometries are the same, and that n is integer, are called into question by emerging structural data.

In mitochondria, the P side is the intermembrane space,

and the N side the mitochondrial matrix; in bacteria, the P side is the outside (the periplasm in gram negative bacteria), the N side the cytoplasm; in chloroplasts, the P side is the lumen and the N side the stroma.

Subunit Composition of the ATP Synthase

There are minor differences between bacteria, mitochondria and chloroplasts in some of the smaller subunits, which leads to a confusing nomenclature. The simplest system is that from *E. coli*. The ATP synthase can be dissociated into two fractions by relatively mild salt treatments.

A soluble portion, the F_1 ATP-ase, contains 5 subunits, in a stoichiometry. Three substrate binding sites are in the subunits. Additional adenine nucleotide binding site in the subunits are regulatory. The F_1 portion catalyses ATP hydrolysis, but not ATP-synthesis.

Dissociation of the F_1 ATP-ase from the membranes of bacteria or organelles leaves behind a membrane embedded portion called F_o. This consists (in *E. coli*) of three subunits a, b and c, with relative stoichiometries of 1:2:9-12. The c-subunit is very hydrophobic, and forms a helix turn helix structure which spans the membrane twice, with a hydrophilic loop on the side of attachment of F_1. There is a conserved acidic residue halfway across the membrane in the C-terminal helix.

After dissociation, the membranes are permeable to protons. The proton leak can be stopped by addition of inhibitors, which are also inhibitors of ATP synthesis in the functional complex. Two "classical" inhibitors are commonly used. Oligomycin binds at the interface between F_o and F_1; dicyclohexylcarbodiimide (DCCD) binds covalently to the conserved acidic residue in the c-subunit of F_o. One DCCD per ATP-ase is sufficient to block turnover, suggesting a cooperative mechanism. The action of these inhibitors indicates that the proton permeability of the F_o is a part of its functional mechanism.

The proton leak can be plugged, and a functional ATP synthase can be reconstituted, by adding back the F_1 portion to membranes containing the F_o portion.

Structure of the F_1 ATP-ase

The structure of the soluble (F_1) portion of the ATP synthase from beef heart mitochondria has been solved by X-ray crystallography.

The protein was crystallised in the presence of ADP, and an ATP analogue, AMP-PNP, in which the two terminal phosphates of ATP were replaced by the non-hydrolysible imidodiphosphate group. The three□ subunits each contained an AMP-PNP. The three□ subunits contained either ADP($□_{DP}$), AMP-PNP($□_{TP}$), or no nucleotide ($□_E$).

Adenosine 5'-triphosphate (ATP) is a multifunctional nucleotide that is most important as a "molecular currency" of intracellular energy transfer. In this role ATP transports chemical energy within cells for metabolism.

It is produced as an energy source during the processes of photosynthesis and cellular respiration and consumed by many enzymes and a multitude of cellular processes including biosynthetic reactions, motility and cell division.

ATP is also incorporated into nucleic acids by polymerases in the processes of DNA replication and transcription. In signal transduction pathways, ATP is used as a substrate by kinases that phosphorylate proteins and lipids, as well as by adenylate cyclase, which uses ATP to produce the second messenger molecule cyclic AMP.

The structure of this molecule consists of a purine base (adenine) attached to the 1' carbon atom of a pentose (ribose). Three phosphate groups are attached at the 5' carbon atom of the pentose sugar. When ATP is used in DNA synthesis, the ribose sugar is first converted to deoxyribose by ribonucleotide reductase.

ATP was discovered in 1929 by Karl Lohmann, and was

proposed to be the main energy-transfer molecule in the cell by Fritz Albert Lipmann in 1941.

Physical and Chemical Properties

ATP consists of adenosine—itself composed of an adenine ring and a ribose sugar—and three phosphate groups (triphosphate). The phosphoryl groups, starting with the group closest to the ribose, are referred to as the alpha (α), beta (β), and gamma (γ) phosphates. ATP is highly soluble in water and is quite stable in solutions between pH 6.8-7.4, but is rapidly hydrolysed at extreme pH, consequently ATP is best stored as an anhydrous salt.

The system of ATP and water under standard conditions and concentrations is extremely rich in chemical energy; the bond between the second and third phosphate groups is loosely said to be particularly high in energy. Strictly speaking, the bond itself is not high in energy (like all chemical bonds it requires energy to break), but energy is produced when the bond is broken and water is allowed to react with the two products.

Thus, energy is produced from the new bonds formed between ADP and water, and between phosphate and water. The net change in enthalpy at Standard Temperature and Pressure of the decomposition of ATP into hydrated ADP and hydrated inorganic phosphate is -20.5 kJ/mole, with a change in free energy of 3.4 kJ/mole. This large release in energy makes the decomposition of ATP in water extremely exergonic, and hence useful as a means for chemically storing energy.

Ionisation in Biological Systems

ATP has multiple ionisable groups with different acid dissociation constants. In neutral solution, ATP is ionised and exists mostly as ATP^{4-}, with a small proportion of ATP^{3-}. As ATP has several negatively-charged groups in neutral solution, it can chelate metals with very high affinity.

The binding constant for various metal ions are (given as per mole) as Mg^{2+} (9 554), Na^{+} (13), Ca^{2+} (3 722), K^{+} (8), Sr^{2+} (1 381) and Li^{+} (25)..Due to the strength of these interactions, ATP exists in the cell mostly in a complex with Mg^{2+}

Figure: Space-filling model of ATP

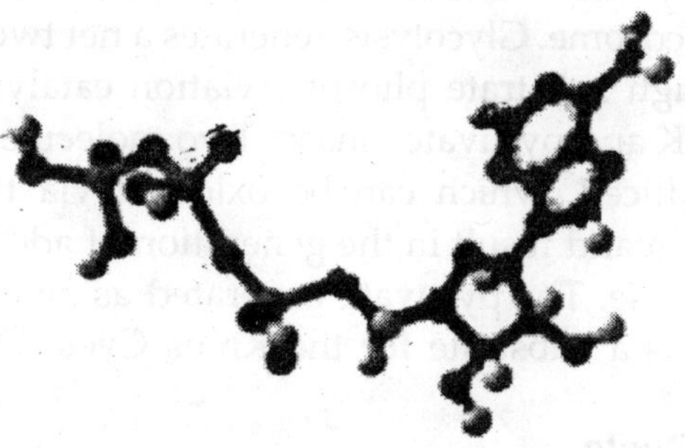

Figure: Ball-and-stick model of ATP

Biosynthesis

ATP can be produced by redox reactions using simple and complex sugars (carbohydrates) or lipids as an energy source. For ATP to be synthesised from complex fuels, they first need to be broken down into their basic components. Carbohydrates are hydrolysed into simple sugars, such as glucose and fructose. Fats (triglycerides) are metabolised to give fatty acids and glycerol.

The overall process of oxidising glucose to carbon dioxide is known as cellular respiration and can produce up to 36 molecules of ATP from a single molecule of glucose. ATP can be produced by a number of distinct cellular processes; the three main pathways used to generate energy in eukaryotic organisms are glycolysis and the citric acid cycle/oxidative phosphorylation, both components of cellular respiration; and beta-oxidation. The majority of this ATP production by a non-photosynthetic aerobic eukaryote takes place in the mitochondria, which can make up nearly 25 per cent of the total volume of a typical cell.

Glycolysis

In glycolysis, glucose and glycerol are metabolised to pyruvate via the glycolytic pathway. In most organisms this process occurs in the cytosol, but in some protozoa such as the kinetoplastids, this is carried out in a specialised organelle called the glycosome. Glycolysis generates a net two molecules of ATP through substrate phosphorylation catalysed by two enzymes: PGK and pyruvate kinase. Two molecules of NADH are also produced, which can be oxidised via the electron transport chain and result in the generation of additional ATP by ATP synthase. The pyruvate generated as an end-product of glycolysis is a substrate for the Krebs Cycle.

Citric Acid Cycle

In the mitochondrion, pyruvate is oxidised by the pyruvate dehydrogenase complex to acetyl CoA, which is fully oxidised to carbon dioxide by the citric acid cycle (also known as the Krebs Cycle). Every "turn" of the citric acid cycle produces two molecules of carbon dioxide, one molecule of the ATP equivalent guanosine triphosphate (GTP) through substrate-level phosphorylation catalysed by succinyl CoA synthetase, three molecules of the reduced co-enzyme NADH, and one molecule of the reduced co-enzyme $FADH_2$. Both of these latter molecules are recycled to their oxidised states (NAD^+

and FAD, respectively) via the electron transport chain, which generates additional ATP by oxidative phosphorylation. The oxidation of an NADH molecule results in the synthesis of about 3 ATP molecules, and the oxidation of one $FADH_2$ yields about 2 ATP molecules. The majority of cellular ATP is generated by this process. Although the citric acid cycle itself does not involve molecular oxygen, it is an obligately aerobic process because O_2 is needed to recycle the reduced NADH and $FADH_2$ to their oxidised states. In the absence of oxygen the citric acid cycle will cease to function due to the lack of available NAD^+ and FAD.

The generation of ATP by the mitochondrion from cytosolic NADH relies on the malate-aspartate shuttle (and to a lesser extent, the glycerol-phosphate shuttle) because the inner mitochondrial membrane is impermeable to NADH and NAD^+. Instead of transferring the generated NADH, a malate dehydrogenase enzyme converts oxaloacetate to malate, which is translocated to the mitochondrial matrix. Another malate dehydrogenase-catalysed reaction occurs in the opposite direction, producing oxaloacetate and NADH from the newly transported malate and the mitochondrion's interior store of NAD^+. A transaminase converts the oxaloacetate to aspartate for transport back across the membrane and into the intermembrane space.

In oxidative phosphorylation, the passage of electrons from NADH and $FADH_2$ through the electron transport chain powers the pumping of protons out of the mitochondrial matrix and into the intermembrane space. This creates a proton motive force that is the net effect of a pH gradient and an electric potential gradient across the inner mitochondrial membrane. Flow of protons down this potential gradient - that is, from the intermembrane space to the matrix - provides the driving force for ATP synthesis by the protein complex ATP synthase. This enzyme contains a rotor subunit that physically rotates relative to the static portions of the protein during ATP synthesis.

Most of the ATP synthesised in the mitochondria will be used for cellular processes in the cytosol; thus it must be exported from its site of synthesis in the mitochondrial matrix. The inner membrane contains antiporters that are integral membrane proteins used to exchange newly-synthesised ATP in the matrix for ADP in the intermembrane space.

Beta-oxidation

Fatty acids can also be broken down to acetyl-CoA by beta-oxidation. Each turn of this cycle reduces the length of the acyl chain by two carbon atoms and produces one NADH and one $FADH_2$ molecule, which are used to generate ATP by oxidative phosphorylation. Because NADH and $FADH_2$ are energy-rich molecules, dozens of ATP molecules can be generated by the beta-oxidation of a single long acyl chain. The high energy yield of this process explains why fat is the best source of dietary calories.

Anaerobic Respiration

Anaerobic respiration or fermentation entails the generation of energy via the process of oxidation in the absence of O_2 as an electron acceptor. In most eukaryotes, glucose is used as both an energy store and an electron donor. The equation for the oxidation of glucose to lactic acid is:

$$C_6H_{12}O_6 \longrightarrow 2CH_3CH(OH)COOH + 2\ ATP$$

In prokaryotes, multiple electron acceptors can be used in anaerobic respiration. These include nitrate, sulphate or carbon dioxide. These processes lead to the ecologically-important processes of denitrification, sulphate reduction and acetogenesis, respectively.

ATP Replenishment by Nucleoside Diphosphate Kinases

ATP can also be synthesised through several so-called "replenishment" reactions catalysed by the enzyme families of nucleoside diphosphate kinases (NDKs), which use other nucleoside triphosphates as a high-energy phosphate donor, and the ATP:guanido-phosphotransferase family, which uses creatine.

$$ADP + GTP \longrightarrow ATP + GDP$$

ATP Production during Photosynthesis

In plants, ATP is synthesised in thylakoid membrane of the chloroplast during the light-dependent reactions of photosynthesis in a process called photophosphorylation. Here, light energy is used to pump protons across the chloroplast membrane. This produces a proton-motive force and this drives the ATP synthase, exactly as in oxidative phosphorylation. Some of the ATP produced in the chloroplasts is consumed in the Calvin cycle, which produces triose sugars.

ATP Recycling

The total quantity of ATP in the human body is about 0.1 mole. The majority of ATP is not usually synthesised *de novo*, but is generated from ADP by the aforementioned processes. Thus, at any given time, the total amount of ATP + ADP remains fairly constant.

The energy used by human cells requires the hydrolysis of 100 to 150 moles of ATP daily which is around 50 to 75 kg. Typically, a human will use up their body weight of ATP over the course of the day. This means that each ATP molecule is recycled 1000 to 1500 times during a single day (100 / 0.1 = 1000). ATP cannot be stored, hence its consumption being followed closely by its synthesis.

Regulation of Biosynthesis

ATP production in an aerobic eukaryotic cell is tightly regulated by allosteric mechanisms, by feedback effects, and by the substrate concentration dependence of individual enzymes within the glycolysis and oxidative phosphorylation pathways. Key control points occur in enzymatic reactions that are so energetically favourable that they are effectively irreversible under physiological conditions.

In glycolysis, hexokinase is directly inhibited by its product, glucose-6-phosphate, and pyruvate kinase is inhibited by ATP itself. The main control point for the glycolytic pathway is phosphofructokinase (PFK), which is allosterically inhibited by high concentrations of ATP and activated by high concentrations of AMP.

The inhibition of PFK by ATP is unusual, since ATP is also a substrate in the reaction catalysed by PFK; the biologically active form of the enzyme is a tetramer that exists in two possible conformations, only one of which binds the second substrate fructose-6-phosphate (F6P). The protein has two binding sites for ATP - the active site is accessible in either protein conformation, but ATP binding to the inhibitor site stabilises the conformation that binds F6P poorly. A number of other small molecules can compensate for the ATP-induced shift in equilibrium conformation and reactivate PFK, including cyclic AMP, ammonium ions, inorganic phosphate, and fructose 1,6 and 2,6 biphosphate.

The citric acid cycle is regulated mainly by the availability of key substrates, particularly the ratio of NAD^+ to NADH and the concentrations of calcium, inorganic phosphate, ATP, ADP, and AMP. Citrate—the molecule that gives its name to the cycle - is a feedback inhibitor of citrate synthase and also inhibits PFK, providing a direct link between the regulation of the citric acid cycle and glycolysis.

In oxidative phosphorylation, the key control point is the reaction catalysed by *cytochrome c oxidase*, which is regulated by the availability of its substrate, the reduced form of *cytochrome c*. The amount of reduced cytochrome c available is directly related to the amounts of other substrates:

$$\frac{1}{2}NADH + cyt\,c_{\alpha x} + ADP + P_i \Leftrightarrow \frac{1}{2}NAD^+ + cyt\,c_{red} + ATP$$

Which directly implies this equation:

$$\frac{cyt\,c_{red}}{cyt\,c_{ox}} = \left(\frac{[NADH]}{[NAD]^{+}}\right)^{\frac{1}{2}} \left(\frac{[ADP][P_i]}{[ATP]}\right) K_{eq}$$

Thus, a high ratio of [NADH] to [NAD^+] or a low ratio of [ADP][P_i] to [ATP] imply a high amount of reduced cytochrome c and a high level of cytochrome c oxidase activity. An additional level of regulation is introduced by the transport rates of ATP and NADH between the mitochondrial matrix and the cytoplasm.

Functions in Cells

ATP is generated in the cell by energy-releasing processes and is broken down by energy-consuming processes, in this way ATP transfers energy between spatially-separate metabolic reactions. ATP is the main energy source for the majority of cellular functions. This includes the synthesis of macromolecules, including DNA, RNA, and proteins. ATP also plays a critical role in the transport of macromolecules across cell membranes, e.g. exocytosis and endocytosis.

ATP is critically involved in maintaining cell structure by facilitating assembly and disassembly of elements of the cytoskeleton. In a related process, ATP is required for the shortening of actin and myosin filament crossbridges required for muscle contraction. This latter process is one of the main energy requirements of animals and is essential for locomotion and respiration.

Cell Signalling

Extracellular Signalling: ATP is also a signalling molecule. ATP, ADP, or adenosine are recognised by purinergic receptors. In humans, this signalling role is important in both the central and peripheral nervous system. Activity-dependent release of ATP from synapses, axons and glia activates purinergic membrane receptors known as P2.

The *P2Y* receptors are *metabotropic*, i.e. G protein-coupled and modulate mainly intracellular calcium and sometimes cyclic AMP levels. Fifteen members of the P2Y family have been reported (P2Y1–P2Y15), although some are only related through weak homology and several (P2Y5, P2Y7, P2Y9, P2Y10) do not function as receptors that raise cytosolic calcium.

The *P2X ionotropic* receptor subgroup comprises seven members (P2X1–P2X7) which are ligand-gated Ca^{2+}-permeable ion channels that open when bound to an extracellular purine nucleotide. In contrast to P2 receptors (agonist order ATP > ADP > AMP > ADO), purinergic nucleotides like ATP are not strong agonists of P1 receptors which are strongly activated by adenosine and other nucleosides (ADO > AMP > ADP > ATP). P1 receptors have A1, A2a, A2b, and A3 subtypes ("A" as a remnant of old nomenclature of *adenosine receptor*), all of which are G protein-coupled receptors, A1 and A3 being coupled to Gi, and A3 being coupled to Gs.

Intracellular Signalling

ATP is critical in signal transduction processes. It is used by kinases as the source of phosphate groups in their phosphate transfer reactions.

Kinase activity on substrates such as proteins or membrane lipids are a common form of signal transduction. Phosphorylation of a protein by a kinase can activate or inhibit the target's activity, these proteins may themselves be kinases, and form part of a signal transduction cascade such as the mitogen-activated protein kinase cascade.

ATP is also used by adenylate cyclase and is transformed to the second messenger molecule cyclic AMP, which is involved in triggering calcium signals by the release of calcium from intracellular stores.

This form of signal transduction is particularly important in brain function, although it is involved in the regulation of a multitude of other cellular processes.

Deoxyribonucleotide Synthesis

In all known organisms, the deoxyribonucleotides that make up DNA are synthesised by the action of ribonucleotide reductase (RNR) enzymes on their corresponding ribonucleotides. This enzyme reduces the 2' hydroxyl group on the ribose sugar to deoxyribose, forming a deoxyribonucleotide (denoted dATP). All ribonucleotide reductase enzymes use a common sulphydryl radical mechanism reliant on reactive cysteine residues that oxidise to form disulphide bonds in the course of the reaction. RNR enzymes are recycled by reaction with thioredoxin or glutaredoxin.

The regulation of RNR and related enzymes maintains a balance of dNTPs relative to each other and relative to NTPs in the cell. Very low dNTP concentration inhibits DNA synthesis and DNA repair and is lethal to the cell, while an abnormal ratio of dNTPs is mutagenic due to the increased likelihood of misincorporating a dNTP during DNA synthesis. Regulation of or differential specificity of RNR has been proposed as a mechanism for alterations in the relative sizes of intracellular dNTP pools under cellular stress such as hypoxia.

Binding to Proteins

An example of the Rossmann fold, a structural domain of a decarboxylase enzyme from the bacterium *Staphylococcus epidermidis* (PDB ID 1G5Q) with a bound flavin mononucleotide co-factor.

Some proteins that bind ATP do so in a characteristic protein fold known as the Rossmann fold, which is a general nucleotide-binding structural domain that can also bind the co-factor NAD. The most common ATP-binding proteins, known as kinases, share a small number of common folds; the protein kinases, the largest kinase superfamily, all share common structural features specialised for ATP binding and phosphate transfer.

ATP in complexes with proteins generally requires the presence of a divalent cation, almost always magnesium, which binds to the ATP phosphate groups. The presence of magnesium greatly decreases the dissociation constant of ATP from its protein binding partner without affecting the ability of the enzyme to catalyse its reaction once the ATP has bound. The presence of magnesium ions can serve as a mechanism for kinase regulation.

ATP Analogs

Biochemistry laboratories often use *in vitro* studies to explore ATP-dependent molecular processes. Enzyme inhibitors of ATP-dependent enzymes such as kinases are needed to experimentally examine the binding sites and transition states involved in ATP-dependent reactions. ATP analogs are also used in X-ray crystallography to determine a protein structure in complex with ATP, often together with other substrates.

Most useful ATP analogs cannot be hydrolysed as ATP would be; instead they trap the enzyme in a structure closely related to the ATP-bound state. Adenosine 5'-(gamma-thiotriphosphate) is an extremely common ATP analog in which one of the gamma-phosphate oxygens is replaced by a sulphur atom; this molecule is hydrolysed at a dramatically slower rate than ATP itself and functions as an inhibitor of ATP-dependent processes. In crystallographic studies, hydrolysis transition states are modelled by the bound vanadate ion. However, caution is warranted in interpreting the results of experiments using ATP analogs, since some enzymes can hydrolyse them at appreciable rates at high concentration.

ATPase

ATPases are a class of enzymes that catalyse the decomposition of adenosine triphosphate (ATP) into adenosine diphosphate (ADP) and a free phosphate ion. This dephosphorylation reaction releases energy, which the enzyme (in most cases) harnesses to drive other chemical reactions that would not otherwise occur. This process is widely used in all known forms of life. Some such enzymes are integral membrane proteins (anchored within biological membranes), and move solutes across the membrane. (These are called *transmembrane ATPases*).

Functions

Transmembrane ATPases import many of the metabolites necessary for cell metabolism and export toxins, wastes, and solutes that can hinder cellular processes. An important example is the sodium-potassium exchanger (or Na^+/K^+ATPase), which establishes the ionic concentration balance that maintains the cell potential. Another example is the hydrogen potassium ATPase (H^+/K^+ATPase or gastric proton pump) that acidifies the contents of the stomach.

Besides exchangers, other categories of transmembrane ATPase include co-transporters and pumps (however, some exchangers are also pumps). Some of these, like the Na^+/K^+ATPase, cause a net flow of charge, but others do not. These are called "electrogenic" and "non-electrogenic" transporters, respectively.

Mechanism

The coupling between ATP hydrolysis and transport is more or less a strict chemical reaction, in which a fixed number of solute molecules are transported for each ATP molecule that is hydrolysed; for example, 3 Na^+ ions out of the cell and 2 K^+ ions inward per ATP hydrolysed, for the Na^+/K^+ exchanger.

Transmembrane ATPases harness the chemical potential energy of ATP, because they perform mechanical work: they transport solutes in a direction opposite to their thermodynamically preferred direction of movement—that is, from the side of the membrane where they are in low concentration to the side where they are in high concentration. This process is considered active transport. For example, the blocking of the vesicular H+-ATPases would increase the pH inside vesicles and decrease the pH of the cytoplasm.

ATP Synthase

The ATP synthase of mitochondria and chloroplasts is an anabolic enzyme that harnesses the energy of a transmembrane proton gradient as an energy source for adding an inorganic phosphate group to a molecule of adenosine diphosphate (ADP) to form a molecule of adenosine triphosphate (ATP).

This enzyme works when a proton moves down the concentration gradient, giving the enzyme a spinning motion. This unique spinning motion bonds ADP and P together to create ATP.

ATP synthase can also function in reverse, that is, used energy released by ATP hydrolysis to pump protons against their thermodynamic gradient.

Phototrophic Bacteria

Cyanobacteria

Cyanobacteria (blue + bacterium) also known as Cyanophyta is a phylum (or "division") of Bacteria that obtain their energy through photosynthesis. They are often still referred to as blue-green algae, although they are in fact prokaryotes like bacteria. The description is primarily used to reflect their appearance and ecological role rather than their evolutionary lineage. Fossil traces of cyanobacteria have been found from around 3.8 billion years ago. They are a major primary producer in the planetary ocean. Their ability to perform oxygenic (plant-like) photosynthesis is thought to have converted the early reducing atmosphere into an oxidising one, which dramatically changed the life forms on Earth and provoked an explosion of biodiversity.

Forms

Cyanobacteria are found in almost every conceivable habitat, from oceans to fresh water to bare rock to soil. Most are found in fresh water, while others are marine, occur in damp soil, or even temporarily moistened rocks in deserts. A few are endosymbionts in lichens, plants, various protists, or sponges and provide energy for the host. Some live in the fur of sloths, providing a form of camouflage.

Cyanobacteria include unicellular and colonial species. Colonies may form filaments, sheets or even hollow balls. Some filamentous colonies show the ability to differentiate into several different cell types: vegetative cells, the normal, photosynthetic cells that are formed under favourable growing conditions; akinetes, the climate-resistant spores that may form when environmental conditions become harsh; and thick-walled heterocysts, which contain the enzyme nitrogenase, vital for nitrogen fixation.

Heterocysts may also form under the appropriate environmental conditions (anoxic) wherever nitrogen is necessary. Heterocyst-forming species are specialised for nitrogen fixation and are able to fix nitrogen gas, which cannot be used by plants, into ammonia (NH_3), nitrites (NO_2^-) or nitrates (NO_3^-), which can be absorbed by plants and converted to protein and nucleic acids. The rice paddies of Asia, which produce about 75 per cent of the world's rice, could not do so were it not for healthy populations of nitrogen-fixing cyanobacteria in the rice paddy fertilizer.

Many cyanobacteria also form motile filaments, called hormogonia, that travel away from the main biomass to bud and form new colonies elsewhere. The cells in a hormogonium are often thinner than in the vegetative state, and the cells on either end of the motile chain may be tapered. In order to break away from the parent colony, a hormogonium often must tear apart a weaker cell in a filament, called a necridium.

Each individual cell of a cyanobacterium typically has a thick, gelatinous cell wall, which stains gram-negative. They lack flagella, but hormogonia and some unicellular species may move about by gliding along surfaces. In water columns some cyanobacteria float by forming gas vesicles, like in archaea.

Photosynthesis

Cyanobacteria have an elaborate and highly organised system of internal membranes which function in photosynthesis. Photosynthesis in cyanobacteria generally uses water as an electron donor and produces oxygen as a by-product, though some may also use hydrogen sulphide as occurs among other photosynthetic bacteria. Carbon dioxide is reduced to form carbohydrates via the Calvin cycle. In most forms the photosynthetic machinery is embedded into folds of the cell membrane, called thylakoids. The large amounts of oxygen in the atmosphere are considered to have been first created by the activities of ancient cyanobacteria. Due to their ability to fix nitrogen in aerobic conditions they are often found as symbionts with a number of other groups of organisms such as fungi (lichens), corals, pteridophytes (Azolla), angiosperms (Gunnera), etc.

Cyanobacteria are the only group of organisms that are able to reduce nitrogen and carbon in aerobic conditions, a fact that may be responsible for their evolutionary and ecological success. The water-oxidising photosynthesis is accomplished by coupling the activity of photosystem (PS) II and I (Z-scheme). In anaerobic conditions, they are also able to use only PS I — cyclic photophosphorylation — with electron donors other than water (hydrogen sulphide, thiosulphate, or even molecular hydrogen) just like purple photosynthetic bacteria. Furthermore, they share an archaebacterial property, which is the ability to reduce elemental sulphur by anaerobic respiration in the dark. Perhaps the most intriguing thing about these organisms is that their photosynthetic electron transport shares the same compartment as the components of respiratory electron transport. Actually, their plasma membrane contains only components of the respiratory chain, while the thylakoid membrane hosts both respiratory and photosynthetic electron transport.

Attached to thylakoid membrane, phycobilisomes act as light harvesting antennae for the photosystems. The phycobilisome components (phycobiliproteins) are responsible for the blue-green pigmentation of most cyanobacteria. The variations to this theme is mainly due to carotenoids and phycoerythrins which give the cells the red-brownish colouration. In some cyanobacteria, the colour of light influences the composition of phycobilisomes. In green light, the cells accumulate more phycoerythrin, whereas in red light they produce more phycocyanin. Thus the bacteria appear green in red light and red in green light. This process is known as complementary chromatic adaptation and is a way for the cells to maximise the use of available light for photosynthesis.

A few genera, however, lack phycobilisomes and have chlorophyll *b* instead (*Prochloron, Prochlorococcus, Prochlorothrix*). These were originally grouped together as the prochlorophytes or chloroxybacteria, but appear to have developed in several different lines of cyanobacteria. For this reason they are now considered as part of cyanobacterial group.

Relationship to Chloroplasts

Chloroplasts found in eukaryotes (algae and higher plants) likely evolved from an endosymbiotic relation with cyanobacteria. This endosymbiotic theory is supported by various structural and genetic similarities. Primary chloroplasts are found among the green plants, where they contain chlorophyll *b*, and among the red algae and glaucophytes, where they contain phycobilins. It now appears that these chloroplasts probably had a single origin, in an ancestor of the clade called Primoplantae. Other algae likely took their chloroplasts from these forms by secondary endosymbiosis or ingestion. It was once thought that the mitochondria in eukaryotes also developed from an endosymbiotic relationship with cyanobacteria; however, we now know that this evolutionary event occurred when aerobic Eubacteria were engulfed by anaerobic host cells. Mitochondria are believed to have originated not from cyanobacteria but from an ancestor of Rickettsia.

Classification

The cyanobacteria were traditionally classified by morphology into five sections, referred to by the numerals I-V. The first three - Chroococcales, Pleurocapsales, and Oscillatoriales - are not supported by phylogenetic studies. However, the latter two - Nostocales and Stigonematales - are monophyletic, and make up the heterocystous cyanobacteria. The members of Chroococales are unicellular and usually aggregated in colonies. The classic taxonomic criterion has been the cell morphology and the plane of cell division. In Pleurocapsales, the cells have the ability to form internal spores (baeocytes). The rest of the sections include filamentous species. In Oscillatorialles, the cells are uniseriately arranged and do not form specialised cells (akinets and heterocysts). In Nostocalles and Stigonematalles the cells have the ability to develop heterocysts in certain conditions. Stigonematales, unlike Nostocalles include species with truly branched trichome. Most taxa included in the phylum or division Cyanobacteria have not yet been validly published under the Bacteriological Code. Except:

- The classes Chroobacteria, Hormogoneae and Gloeobacteria
- The orders Chroococcales, Gloeobacterales, Nostocales, Oscillatoriales, Pleurocapsales and Stigonematales
- The families Prochloraceae and Prochlorotrichaceae
- The genera Halospirulina, Planktothricoides, Prochlorococcus, Prochloron, Prochlorothrix.

Biotechnology and Applications

Certain cyanobacteria produce cyanotoxins like Anatoxin-a, Anatoxin-as, Aplysiatoxin, Cylindrospermopsin, Domoic acid, Microcystin LR, Nodularin R (from *Nodularia*), or Saxitoxin. Sometimes a mass-reproduction of cyanobacteria results in algal blooms.

The unicellular cyanobacterium *Synechocystis* sp. PCC 6803

was the first photosynthetic organism whose genome was completely sequenced (in 1996, by the Kazusa Research Institute, Japan). It continues to be an important model organism. Some cyanobacteria are sold as food, notably *Aphanizomenon flosaquae* and *Arthrospira platensis* (Spirulina). It has been suggested that they could be a much more substantial part of human food supplies, as a kind of superfood.

Along with algae, some hydrogen producing cyanobacteria are being considered as an alternative energy source, notably at Oregon State University, in research supported by the US Department of Energy, Princeton University, Colorado School of Mines as well as at Uppsala University, Sweden.

Health Risks

Some species of cyanobacteria produce neurotoxins, hepatotoxins, cytotoxins, and endotoxins, making them dangerous to animals and humans. Several cases of human poisoning have been documented but a lack of knowledge prevents an accurate assessment of the risks.

Oxygen Catastrophe

The Oxygen Catastrophe was a massive environmental change believed to have happened during the Siderian period at the beginning of the Paleoproterozoic era, about 2.4 billion years ago. The change was catastrophic to the organisms living at the time because there was an increase in oxygen which was toxic to them. It is also called the *Oxygen Crisis, Oxygen Revolution* or *The Great Oxidation.*

When evolving life forms developed oxyphotosynthesis about 2.7 billion years ago, molecular oxygen was produced in large quantities. This means there was a time lag of about 300 million years between the time oxygen production from photosynthetic organisms started, and the Oxygen Catastrophe.

One phenomenon that explains this lag is that oxidation had to await tectonically driven changes in Earth's 'anatomy', including the appearance of shelf seas where reduced organic carbon could reach the sediments and be buried. The newly-produced oxygen was also tied up in chemical reactions in the oceans, primarily with iron. Evidence for this is found in older rocks that contain massive banded iron formations that were apparently laid down as iron and oxygen first combined. But these phenomena do not seem to account for the lag completely.

Photosynthetic organisms were also a source of methane. This was also a big trap for molecular oxygen, because methane oxidises readily in the presence of UV radiation. A new idea to explain the 300 million year lag comes from a mathematical model of the atmosphere. The result is a system with two steady states with lower (0.02 per cent) and higher (21 per cent or more) content of oxygen. The Great Oxidation can be understood as a switch between lower and upper stable steady states. This bistability was caused by UV shielding decreasing the rate of methane oxidation once oxygen levels were sufficient to form an ozone layer. Another part of the effect may have been photosynthetic production of molecular hydrogen which as it formed got into the atmosphere and was slowly lost to space. The plentiful oxygen eventually caused an ecological crisis. From the point of the view of the anaerobic organisms already present on earth and in the oceans, the environment was poisoned with oxygen, and the oceans were scrubbed free of important nutrients. It is from their point of view, and because of the drastic change that the oxygen caused to the air and oceans, that the Oxygen Catastrophe gets its name.

However, it also provided a new opportunity. Despite recycling, life had remained energetically limited until the widespread availability of oxygen. This breakthrough in metabolic evolution greatly increased the free energy supply to living organisms, having a truly global environmental impact.

Allophycocyanin

Allophycocyanin is a protein from the light-harvesting phycobiliprotein family, along with phycocyanin, phycoerythrin and phycoerythrocyanin. It is an accessory pigment to chlorophyll. All phycobiliproteins are water soluble and therefore cannot exist within the membrane like carotenoids, but aggregate forming clusters that adhere to the membrane called phycobilisomes. Allophycocyanin absorbs and emits red light (650 & 660 nm max, respectively), and is readily found in Cyanobacteria (also called blue-green algae), and red algae. Phycobilin pigments have fluorescent properties that are used in immunoassay kits. In flow cytometry, it is often abbreviated APC.

Anabaena

Anabaena is a genus of filamentous cyanobacteria, or blue-green algae, found as plankton. It is known for its nitrogen fixing abilities, and they form symbiotic relationships with certain plants, such as the mosquito fern. They are one of four genera of cyanobacteria that produce neurotoxins, which are harmful to local wildlife, as well as farm animals and pets.

A DNA sequencing project was undertaken in 1999, which mapped the complete genome of *Anabaena*, which is 7.2 million base pairs long. Certain species of *Anabaena* have been used on rice paddy fields, proving to be an effective natural fertilizer.

Species

- *Anabaena aequalis*
- *Anabaena affinis*
- *Anabaena angstumalis*
  — *Anabaena angstumalis angstumalis*
  — *Anabaena angstumalis marchica*
- *Anabaena aphanizomendoides*
- *Anabaena azollae*
- *Anabaena bornetiana*
- *Anabaena catenula*
- *Anabaena circinalis*

- *Anabaena confervoides*
- *Anabaena constricta*
- *Anabaena cycadeae*
- *Anabaena cylindrica*
- *Anabaena echinispora*
- *Anabaena felisii*
- *Anabaena flosaquae*
  - — *Anabaena flosaquae flosaquae*
  - — *Anabaena flosaquae minor*
  - — *Anabaena flosaquae treleasei*
- *Anabaena helicoidea*
- *Anabaena inaequalis*
- *Anabaena lapponica*
- *Anabaena laxa*
- *Anabaena lemmermannii*
- *Anabaena levanderi*
- *Anabaena limnetica*
- *Anabaena macrospora*
  - — *Anabaena macrospora macrospora*
  - — *Anabaena macrospora robusta*
- *Anabaena monticulosa*
- *Anabaena oscillarioides*
- *Anabaena planctonica*
- *Anabaena raciborskii*
- *Anabaena scheremetievi*
- *Anabaena sphaerica*
- *Anabaena spiroides*
  - — *Anabaena spiroides crassa*
  - — *Anabaena spiroides spiroides*
- *Anabaena subcylindrica*
- *Anabaena torulosa*

- *Anabaena unispora*
- *Anabaena variabilis*
- *Anabaena verrucosa*
- *Anabaena viguieri*
- *Anabaena wisconsinense*
- *Anabaena zierlingii*

Fat Choy

Fat choy (*Nostoc flagelliforme*), also known as *black moss, faat choy* or *hair moss,* is a terrestrial cyanobacterium (a type of photosynthetic bacteria) that is used as a vegetable in Chinese cuisine.

When dried, the product has the appearance of black hair. For that reason, its name in Chinese means "hair vegetable". The last two syllables of this name in Cantonese sound the same as another Cantonese saying meaning "struck it rich" (though the second syllable, *choi,* has a different tone) — this is found, for example, in the Cantonese saying, "*Gung hei faat choi*" (meaning "congratulations and be prosperous"), which is often proclaimed during Chinese New Year. For that reason, this product is a popular ingredient in dishes used for the Chinese New Year. It is enjoyed for its texture, which is like very fine vermicelli. Fat choy grows on the ground in the Gobi Desert and the Qinghai plateau. Over-harvesting on the Mongolian steppes has furthered erosion and desertification in those areas. The Chinese government has limited its harvesting, which has caused its price to increase. This may be one reason why some commercially available fat choy has been found to be adulterated with strands of a non-cellular starchy material, with other additives and dyes. Real fat choy is dark green in colour, while the counterfeit fat choy appears black.

Health Effects

A research team from the biochemistry department of the Chinese University of Hong Kong said that international research has shown that fat choy, besides having no nutritional value, has also been found to contain Beta-methylamino L-alanine (BMAA), a toxic amino acid that could affect the normal functions of nerve cells. Professor Chan King-ming of the team told the media that eating fat choy could lead to degenerative diseases such as Alzheimer's, Parkinson's, and dementia.

Gloeocapsa

Gloeocapsa is a type of Photoautotrophic Bacteria (Cyanobacteria), and is a prokaryote. The cells secrete individual gelatinous sheaths which can often be seen as sheaths around recently divided cells within outer sheaths. Recently divided cell pairs often appear to be only one cell since the new cells cohere temporarily.

Heterocyst

Heterocysts are specialised nitrogen-fixing cells formed by some filamentous cyanobacteria, such as *Nostoc punctiforme, Cylindrospermum stagnale* and *Anabaena sperica*, during nitrogen starvation. They fix nitrogen from dinitrogen (N_2) in the air using the enzyme nitrogenase, in order to provide the cells in the filament with nitrogen for biosynthesis. Nitrogenase is inactivated by oxygen, so the heterocyst must create a microanaerobic environment. The heterocysts' unique structure and physiology requires a global change in gene expression. For example, heterocysts:

- Produce three additional cell walls, including one of glycolipid that forms a hydrophobic barrier to oxygen
- Produce nitrogenase and other proteins involved in nitrogen fixation
- Degrade photosystem II, which produces oxygen
- Up-regulate glycolytic enzymes, which use up oxygen and provide energy for nitrogenase
- Produce proteins that scavenge any remaining oxygen.

Cyanobacteria usually obtain a fixed carbon (carbohydrate)

by photosynthesis. The lack of photosystem II prevents heterocysts from photosynthesising, so the vegetative cells provide them with carbohydrates, which is thought to be sucrose. The fixed carbon and nitrogen sources are exchanged though channels between the cells in the filament. Heterocysts maintain photosystem I, allowing them to generate ATP by cyclic photophosphorylation.

Single heterocysts develop about every 9-15 cells, producing a one-dimensional pattern along the filament. The interval between heterocysts remains approximately constant even though the cells in the filament are dividing. The bacterial filament can be seen as a multicellular organism with two distinct yet interdependent cell types. Such behaviour is highly unusual in prokaryotes and may have been the first example of multicellular patterning in evolution.

Once a heterocyst has formed, it cannot revert to a vegetative cell, so this differentiation can be seen as a form of apoptosis. Certain heterocyst-forming bacteria can differentiate into spore-like cells called akinetes or motile cells called hormogonia, making them the most phenotyptically versatile of all prokaryotes.

The mechanism of controlling heterocysts is thought to involve the diffusion of an inhibitor of differentiation called pats. Heterocyst formation is inhibited in the presence of a fixed nitrogen source, such as ammonium or nitrate. Heterocyst maintenance is dependent on an enzyme called hetN. The bacteria may also enter a symbiotic relationship with certain plants.

In such a relationship, the bacteria do not respond to the availability of nitrogen, but to signals produced by the plant. Up to 60 per cent of the cells can become heterocysts, providing fixed nitrogen to the plant in return for fixed carbon.

The cyanobacteria that form heterocysts are divided into the orders Nostocales and Stigonematales, which form simple and branching filaments respectively.

Hormogonales

The term *"Hormogonales"* is no longer in use. It once indicated an order inside the class *Cyanobacteria* comprising the former suborders *"Nostocales"*, *"Scytonematales"* and *Stigonematales*. Some species belonging to *Hormogonales* would be *Anabaena, Nostoc, Rivularia, Scytonema, Stigonema*.

Lyngbya Majuscula

Lyngbya majuscula is a species in the genus Lyngbya (a seaweed). It is a cyanobacteria that is one of the causes of the human skin irritation *seaweed dermatitis*. It is also known as Fireweed.

"Cyanobacteria are an ancient and diverse group of photosynthetic microorganisms, which inhabit many different and extreme environments. This indicates a high degree of biological adaptation, which has enabled these organisms to thrive and compete effectively in nature. The filamentous cyanobacterium, Lyngbya majuscula, produces several promising anti-fungal and cytotoxic agents, including laxaphycin A and B and curacin A."

This organism appears on the increase due to pollution and overfishing. Nutrients such as nitrogen and human waste flow to the ocean due to rain runoff and sewers. These nutrients increase the population of microbes, which in turn remove oxygen from the water. Lack of fish to eat the microbes furthers the microbe populations. Low oxygen is the environment that Cyanobacteria evolved for.

Nodularia

Nodularia is a genus of filamentous cyanobacteria, or blue-green algae. They occur mainly in brackish or slightly salinic waters, e.g. in the Baltic Sea. Nodularia cells occasionally form heavy algal blooms. Some strains produce a toxin called Nodularin R, which is harmful to humans.

Nostoc

Nostoc is a genus of fresh water cyanobacteria that forms spherical colonies composed of filaments of moniliform cells in a gelatinous sheath. When on the ground, a Nostoc colony is ordinarily not séen; but after a rain it swells up into a conspicuous jelly-like mass, which was once thought to have fallen from the sky, whence the popular names, *fallen star* and *star jelly*. It is also called *witches' butter*.

It can be found on moist rocks, at the bottom of lakes and springs, and rarely in marine habitats. These bacteria sometimes contain photosynthetic pigments in their cytoplasm to perform photosynthesis. It may also grow symbiotically within the tissues of plants, such as the aquatic fern *Azolla* (mosquito fern) or hornworts, providing nitrogen to its host.

Species

Nostoc is a member of the family Nostocaceae of the order Hormogonales. Species include:

- *N. azollae*
- *N. caeruleum*
- *N. carneum*
- *N. comminutum*
- *N. commune*
- *N. ellipsosporum*
- *N. flagelliforme*
- *N. linckia*
- *N. longstaffi*
- *N. microscopicum*
- *N. muscorum*
- *N. paludosum*
- *N. pruniforme*
- *N. punctiforme*

- *N. sphaericum*
- *N. spongiaeforme*
- *N. verrucosum*

Culinary Use

Containing protein and vitamin C, *Nostoc* species are cultivated and consumed as a foodstuff, primarily in Asia. The *N. flagelliforme* and *N. commune* varieties are consumed in China, Japan and Java. The preferred variety in Central Asia is *N. ellipsosporum*.

Nostocaceae

The Nostocaceae is a family of cyanobacteria that forms filament-shaped colonies enclosed in mucus or a gelatinous sheath. Some genera in this family are found primarily in fresh water (such as *Nostoc*), while others are found primarily in salt water (such as *Nodularia*). Other genera (e.g. *Anabaena*) may be found in both fresh and salt water. Like other cyanobacteria, these bacteria sometimes contain photosynthetic pigments in their cytoplasm to perform photosynthesis. The particular pigments they contain gives the cells a bluish-green colour.

Species of the Nostocaceae are particularly known for their nitrogen fixing abilities, and they form symbiotic relationships with certain plants, such as the mosquito fern, cycads, and hornworts. The cyanobacteria provides nitrogen to its host. Certain species of *Anabaena* have been used on rice paddy fields. Mosquito ferns carrying the cyanobacteria grow on the water in the fields during the growing season. They and the nitrogen they contain are then ploughed into the soil following the harvest, which has proved to be an effective natural fertilizer.

The family Nostocaceae belongs to the order Nostocales. Members of the family can be distinguished from those in other families by their (1) unbranched filaments of cells arranged end-to-end, and (2) development of heterocysts among the cells of the filaments.

Oscillatoria

Oscillatoria is a genus of filamentous bacteria which is named for the oscillation in its movement (cyanobacteria). It is commonly found in watering-troughs waters, and is mainly blue-green or brown-green. Oscillatoria is an organism that reproduces by fragmentation. Oscillatoria forms long filaments of cells which can break into fragments called hormongia. The hormogonia can grow into a new, longer filament. Breaks in the filament usually occur where dead cells are present.

Photosynthetic Picoplankton

Photosynthetic picoplankton is the fraction of the plankton performing photosynthesis composed by cells between 0.2 and 2 μm (picoplankton). It is especially important in the central oligotrophic regions of the world oceans that have very low concentration of nutrients.

History

- 1952: Description of the first really picoplanktonic species, *Chromulina pusilla*, by Butcher. This species was renamed in 1960 *Micromonas pusilla* and is now known as one of the most abundant in temperate oceanic waters.
- 1979 : Discovery of marine *Synechococcus* by Waterbury and confirmation with electron microscopy by Johnson and Sieburth.
- 1982 : The same Johnson and Sieburth demonstrate the importance of small eukaryotes by electron microscopy.
- 1983 : W.K. Li and Platt show that a large fraction of marine primary production is due to organisms smaller than 2 μm.
- 1986 : Discovery of "prochlorophytes" by Chisholm and Olson in the Sargasso Sea, baptised in 1992 *Prochlorococcus marinus*.
- 1994 : Discovery in the Thau lagoon in France of the

smallest photosynthetic eukaryote known to date, *Ostreococcus tauri*, by Courties.

- 2001 : Through sequencing of the ribosomal RNA gene extracted from marine samples, several European teams discover simultaneously that eukaryotic picoplankton is highly diversified.

Methods of Study

Because of its very small size, picoplankton is difficult to study by classic methods such as optical microscopy. More sophisiticated methods are needed.

- Epifluorescence microscopy allows to detect certain groups of cells possessing fluorescent pigments such as *Synechococcus* which possess phycoerythrin.
- Flow cytometry measures the size (" side scatter ") and fluorescence on 1,000 in 10,000 cells per second. It allows one to determine very easily the concentration of the various picoplankton populations on marine samples. Three groups of cells (*Prochlorococcus*, *Synechococcus* and picoeukaryotes) can be distinguished. For example *Synechococcus* is characterised by the double fluorescence of its pigments: orange for phycoerythrin and red for chlorophyll. Flow cytometry also allows to sort out specific populations (for example *Synechococcus*) in order put them in culture, or to make more detailed analyses.
- Analysis of photosynthetic pigments such as chlorophyll or carotenoids by high precision chromatography (HPLC) allows to determine the various groups of algae present in a sample.
- Molecular biology techniques:
  - — Cloning and sequencing of genes such as that of ribosomal RNA, which allows to determine total diversity within a sample.
  - — DGGE (Denaturing Gel Electrophoresis), that is

faster than the previous approach allows to have an idea of the global diversity within a sample.

— In situ hybridisation (FISH) uses fluorescent probes recognising specific taxon, for example a species, a genus or a class.

— Real-time PCR can be used, as FISH, to determine, the abundance of specific groups. It has the main advantage to allow the rapid analysis of a large number of samples simultaneously, but requires more sophisticated controls and calibrations.

Composition

Three major groups of organisms constitute photosynthetic picoplankton.

- Cyanobacteria belonging to the genus *Synechococcus* of a size of 1 μm (micrometre) were first discovered in 1979 by J. Waterbury (Woods Hole Oceanographic Institution). They are quite ubiquist, but most abundant in relatively mesotrophic waters.

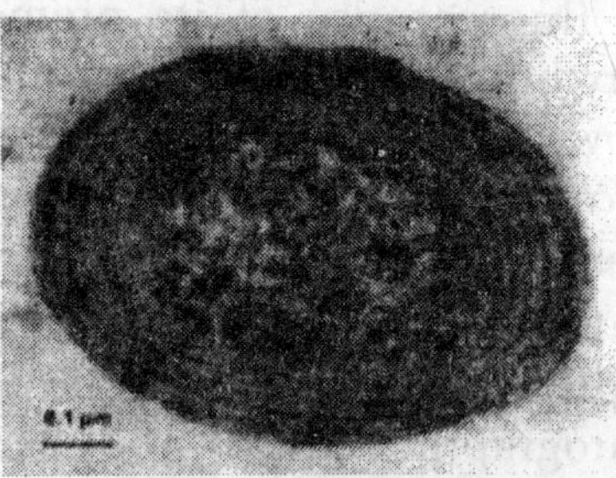

Figure: Prochlorococcus

- Cyanobacteria belonging to the genus *Prochlorococcus* are particularly remarkable. With a typical size of 0.6 μm, *Prochlorococcus* was discovered only in 1988 by two American researchers, Sallie W. (Penny) Chisholm (Massachusetts Institute of Technology) and R.J. Olson (Woods Hole Oceanographic Institution). In spite of its small size, this photosynthetic organism is undoubtedly the most abundant of the planet: indeed its density can reach up to 100 million cells per litre and it can be found down to a depth of 150 m in all the intertropical belt.

- Picoplanktonic eukaryotes are the least well known, as demonstrated by the recent discovery of major groups. Andersen created in 1993 a new class of brown algae, the Pelagophyceae. More surprising still, the discovery in 1994 of an eukaryote of very small size, *Ostreococcus tauri*, dominating the phytoplanktonic biomass of a French brackish lagoon (etang de Thau), shows that these organisms can also play a major ecological role in coastal environments. In 1999, yet a new class of alga was discovered, the Bolidophyceae, very close genetically of diatoms, but quite different morphologically. At the present time, about 50 species are known belonging to several classes.

Algal classes containing picoplankton species

Classes	***Picoplanktonic genera***
Chlorophyceae	*Nannochloris*
Prasinophyceae	*Micromonas , Ostreococcus, Pycnococcus*
Prymnesiophyceae	*Imantonia*
Pelagophyceae	*Pelagomonas*
Bolidophyceae	*Bolidomonas*
Dictyochophyceae	*Florenciella*

In Situ Diversity

The introduction of the molecular biology in oceanography revolutionised the knowledge of the oceanic ecosystems. For the first time, we were able to determine the composition of the picoplanktonic compartment without having neither to observe it, nor to cultivate him. In practice, it is possible to determine the sequence of a gene present in all living organisms, the one coding the small sub-unit of the ribosomal RNA (rRNA).

Every species has a sequence which is specific and two closely related species (for example, the man and the chimpanzee) have very similar sequences. The analysis of the sequence allows to place an organism in the phylogenetic tree of the life. Furthermore we can determine on this gene small regions characteristics of a group of organisms (for example a specific genus of the picoplankton such as "Ostreococcus") and synthetise a "probe" recognising this region. If this probe is marked with a fluorescent compound and put in contact with cells, only cells belonging to the targeted group will be visible under a fluorescence microscope (FISH technique).

These approaches implemented since the 1990s for bacteria, were applied to the photosynthetic picoeukaryotes only 10 years later. They revealed a very wide diversity and put in light the importance of the following groups in the picoplankton :

- Prasinophyceae
- Haptophyta
- Cryptophyta

In temperate coastal environment, the genus *Micromonas* (Prasinophyceae) seems dominant. However, in numerous oceanic environments, the dominant species of eukaryotic picoplankton remain still unknown.

Ecology

Each picoplanktonic population occupies a specific ecological niche in the oceanic environment.

- The *Synechococcus* cyanobacterium is generally abundant in mesotrophic environments, for example in the vicinity of the equatorial upwelling or in coastal regions.
- The *Prochlorococcus* cyanobacterium replaces it when the waters becomes impoverished in nutrients (i.e. oligotrophic). On the other hand in temperate

region (for example in the North Atlantic Ocean), *Prochlorococcus* is absent because the cold waters prevent its development.

- The diversity of eukaryotes, corresponds undoubtedly to a big variety of environments. In oceanic regions, they are often observed at depth at the base of the well-lit layer (the "euphotic" layer). In coastal regions, certain sorts of picoeukaryotes such as "Micromonas" dominate. Their abundance follows a seasonal cycle, as the plankton of bigger size, with a maximum in summer.

Thirty years ago, it was hypothesised that the speed of division for microorganisms in central oceanic ecosystems was very slow, of the order of one week or one month. This hypothesis was consolidated by the fact that the biomass (estimated for example by the contents of chlorophyll) was very stable over time. However with the discovery of the picoplankton, it was found that the system was much more dynamic than previously thought. In particular, small predators of a size of a few microns which ingest picoplanktonic algae as quickly as they were produced, were found to be ubiquitous. This extremely sophisticated predator-prey system is practically always at equilibrium and results in a quasi-constant picoplankton biomass. This perfect equivalence between production and consumption makes it however extremely difficult to measure precisely the speed at which the system turns over.

In 1988, two American researchers, Carpenter and Chang, had suggested estimating the speed of cell division of phytoplankton by following the course of DNA replication by microscopy. By replacing the microscope by a flow cytometer, it is possible to follow the DNA content of picoplankton cells (for example *Prochlorococcus*) over time. This allowed to establish that picoplankton cells are extremely synchronous: they replicate their DNA and then divide all at the same time at the end of the day. This synchronisation could be due to the presence of a biological internal clock.

Genomics

In the 2000s, genomics allowed to cross a supplementary stage. Genomics consists in determining the complete sequence of genome of an organism and to list every gene present. It is then possible to get an idea of the metabolic capacities of the targeted organisms and understand how it adapts to its environment. To date, the genomes of several types of *Prochlorococcus* and *Synechococcus*, and of a strain of *Ostreococcus* have been determined, while those of several other cyanobacteria and of small eukaryotes (*Bathycoccus*, *Micromonas*) are under sequencing. In parallel, genome analyses begin to be done directly from oceanic samples (ecogenomics or metagenomics), allowing us to access to large sets of gene for uncultivated organisms.

Genomes of photosynthetic picoplankton strains that have been sequenced to date

Genus	*Strain*	*Sequencing centre*	*Remark*
Prochlorococcus	MED4	JGI	
	SS120	Genoscope	
	MIT9312	JGI	
	MIT9313	JGI	
	NATL2A	JGI	
	CC9605	JGI	
	CC9901	JGI	
Synechococcus	WH8102	JGI	
	WH7803	Genoscope	
	$RCC_{3}07$	Genoscope	
	CC9311	TIGR	
Ostreococcus	OTTH95	Genoscope	

Prochlorococcus

Prochlorococcus is a genus of very small (0.6 μm) marine cyanobacteria with an unusal pigmentation (chlorophyll *b*) belonging to photosynthetic picoplankton. It is probably the most abundant photosynthesis organism on Earth.

Although there had been several earlier records of very small chlorophyll-*b*-containing cyanobacteria in the ocean,

Prochlorococcus was actually discovered in 1986 by Sallie W. (Penny) Chisholm of the Massachusetts Institute of Technology, Robert J. Olson of the Woods Hole Oceanographic Institution, and other collaborators in the Sargasso Sea using flow cytometry. The first culture of *Prochlorococcus* was isolated in the Sargasso Sea in 1988 (strain SS120) and shortly another strain was obtained from the Mediterranean Sea (strain MED). The name *Prochlorococcus* originated from the fact it was originally assumed that *Prochlorococcus* was related to *Prochloron* and other chlorophyll *b* containing bacteria, called prochlorophytes, but it is now known that prochlorophytes form several separate phylogenetic groups within the cyanobacteria subgroup of the bacteria kingdom.

Marine cyanobacteria are to date the smallest known photosynthetic organisms: *Prochlorococcus* is the smallest at just 0.5 to 0.8 micrometres across. Possibly they are also the most plentiful species on Earth: a single ml of surface seawater may contain 100,000 cells or more. Worldwide, there are estimated to be 100 octillion individuals. *Prochlorococcus* is ubiquitous between 40°N and 40°S and dominates in the oligotrophic (nutrient poor) regions of the oceans.

The light harvesting pigment complement of *Prochlorococcus* is unique, consisting predominantly of divinyl derivatives of chlorophyll *a* (Chl a_2) and *b* (Chl b_2) and lacking monovinyl chlorophylls. *Prochlorococcus* occupies two distinct niches, leading to the nomenclature of the low light (LL) and high light (HL) groups, which vary in pigment ratios (LL possess a high ration of chlorophyll b_2: a_2 and HL low b_2: a_2), light requirements, nitrogen and phosphorus utilisation, copper and virus sensitivity. These "ecotypes" can be differentiated on the basis of the sequence of their ribosomal RNA gene. Recently the genomes of several strains of *Prochlorococcus* have been sequenced.

Stromatolite

Stromatolites (from Greek *stroma*, mattress, bed, stratum,

and *lithos*, rock) are defined as "attached, lithified sedimentary growth structures, accretionary away from a point or limited surface of initiation." A variety of stromatolite morphologies exist including conical, stratiform, branching, domal, and columnar types. Stromatolites are commonly thought to have been formed by the trapping, binding, and cementation of sedimentary grains by microorganisms, especially cyanobacteria (formerly known as blue-green algae). However, very few ancient stromatolites actually contain fossilised microbes. While features of some stromatolites are suggestive of biological activity, others possess features that are more consistent with "abiotic" (non-organic) precipitation. Finding reliable ways to distinguish between biologically-formed and abiotic (non-biological) "stromatolites" is an active area of research in geology.

Stromatolites were much more abundant on the planet in Pre-cambrian times. While older, Archean (a.k.a. Archeozoic) fossil remains are presumed to be single-celled colonies of blue-green bacteria, younger (that is, Proterozoic) fossils may be primordial forms of the eukaryote chlorophytes (that is, green algae). One genus of stromatolite very common in the geologic record is *Collenia*.

Prior to 2.4 billion years ago, the earth's atmosphere was rich in carbon dioxide. However, the Pre-cambrian air lacked the oxygen that sustains the complex multicellular life that has evolved since the "Cambrian explosion" 540 million years ago. Stromatolites in the fossil record decline sharply in both diversity and number during the late Proterozoic eon, although they are present, but not common, in Paleozoic era strata. Today, stromatolites are quite uncommon in marine environments. As a result, they have become valuable "living fossils."

Their former abundance may be because there were no grazing animals back during the Pre-cambrian to destabilise sediments and consume growing microbial mats, thereby favouring the preservation of these microbialites. Also, changing chemical conditions in the ocean during this time could be responsible for the precipitation of non-biological stromatolites through the growth of tiny crystals.

While prokaryotic cyanobacteria themselves reproduce asexually through cell division, they were instrumental in priming the environment for the evolutionary development of more complex eukaryotic organisms. Cyanobacteria are thought to be largely responsible for increasing the amount of oxygen in the primeval earth's atmosphere through their continuing photosynthesis.

Cyanobacteria use water, carbon dioxide, and sunlight to create their food. The by-products of this process are oxygen and calcium carbonate (lime). A layer of mucous often forms over mats of cyanobacterial cells. In modern microbial mats, debris from the surrounding habitat can become trapped within the mucous, which can be cemented together by the calcium carbonate to grow thin laminations of limestone. These laminations can accrete over time, resulting in the banded pattern common to stromatolites. The domal morphology of biological stromatolites is the result of the vertical growth necessary for the continued infiltration of sunlight to the organisms for photosynthesis.

Modern stromatolites are mostly found in hypersaline lakes and marine lagoons where extreme conditions exclude animal grazing. One such location is Hamelin Pool Marine Nature Reserve, Shark Bay in Western Australia where excellent specimens are today observed. Freshwater stromatolites can be found in Cuatro Cienegas, a unique ecosystem in the Mexican desert.

Layered spherical growth structures *similar* to stromatolites, named "oncolites," are also known from the fossil record.

Synechococcus

Synechococcus is a unicellular cyanobacterium that is very wide spread in the marine environment. Its size varies between 0.8 and 1.5 µnm. The photosynthetic coccoid cells are preferentially found in surface well lit waters where this bacterium can be very abundant (generally from 1,000 to 200,000 of cells by millilitre).

The genome of *Synechococcus elongatus* strain PCC7002 has a size of 2,7 Mpb, that of the oceanic strain WH8102 is 2.4 Mbp.

Synechococcus is one of the most important component of the prokaryotic autotrophic picoplankton in the temperate to tropical oceans. The organisms was first described 1979 (Johnson & Sieburth 1979, Waterbury *et al.* 1979) and was originally defined to include "small unicellular cyanobacteria with ovoid to cylindrical cells that reproduce by binary traverse fission in a single plane and lack sheaths" (Rippka *et al.* 1979). This definition of the genus *Synechococcus* contained organisms of considerable genetic diversity and was later subdivided into subgroups based on the presence of the accessory pigment phycoerythrin. The marine forms of *Synechococcus* are coccoid cells between 0.6 and 1.6 µm in size. They are gram negative cells with highly structured cell walls that may contain projections on their surface (Perkins *et al.* 1981). Electron microscopy frequently reveals the presence of phosphate inclusions, glycogen granules and more importantly highly structured carboxysomes.

Cells are known to be motile by a gliding type method (Castenholz 1982) and a novel uncharacterised, non-phototactic swimming method (Waterbury *et al.* 1985) that does not involve flagellar motion. While some cyanobacteria are capable of photoheterotrophic or even chemoheterotrophic growth, all marine *Synechococcus* strains appear to be obligate photoautotrophs (Waterbury *et al.* 1986b) that are capable of supporting their nitrogen requirements using nitrate, ammonia or in some cases urea as a sole nitrogen source. Marine *Synechococcus* are traditionally not thought to fix nitrogen.

Pigments

The main photosynthetic pigment in *Synechococcus* in chlorophyll a, while its major accessory pigments are phycobillin proteins (Waterbury *et al.* 1979). The four commonly recognised phycobillins are phycocyanin, allophycocyanin, allophycocyanin B and phytoerythrin (Stanier & Cohen-Bazire 1977). In addition *Synechococcus* also contains zeaxanthin but no diagnostic pigment for this organism is known. Zeaxanthin is also found in Prochlorococcus, rhodophytes and as a minor pigment in some chlorophytes and eustigmatophytes. Similarly phycoerythrin is also found in rhodophytes and some cryptomonads (Waterbury *et al.* 1986b).

Phylogeny

Phylogenetic description of *Synechococcus* is difficult. Isolates are morphologically very similar, yet exhibited a G+C content ranging between 39 and 71 per cent (Waterbury *et al.* 1986b) illustrating the large genetic diversity of this provisional taxon. Initially attempts were made to divide the group into three sub-clusters, each with a specific range of genomic G+C content (Rippka & Cohen-Bazire 1983). The observation that open-ocean isolates alone nearly span the complete G+C spectrum however indicates that *Synechococcus* is composed of at least several species. Bergey's Manual (Herdman *et al.* 2001) now divides Synechococcus into five clusters (equivalent to genera) based on morphology, physiology and genetic traits.

Cluster one includes relatively large (1-1.5μm) non-motile obligate photoautotrophs that exhibit low salt tolerance. Reference strains for this cluster are PCC6301 (formerly *Anacycstis nidulans*) and PCC6312, which were isolated from freshwater in Texas and California respectively (Rippka *et al.* 1979). Cluster 2 also is characterised by low salt tolerance. Cells are obligate photoautrotrophs, lack phycoerythrin and are thermophilic. The reference strain PCC6715 was isolated from a hot spring in Yellowstone National Park (Dyer & Gafford 1961). Cluster 3 includes phycoerythrin lacking marine *Synechococcus* that are euryhaline, i.e. capable of growth in both marine and fresh water environments. Several strains, including the reference strain PCC7003 are facultative heterotrophs and require vitamin B12 for growth. Cluster 4 contains a single isolate, PCC7335. This strain is obligate marine (Waterbury & Stanier 1981). This strain contains phycoerthrin and was first isolated from the intertidal zone in Perto Penasco, Mexico (Rippka *et al.* 1979). The last cluster contains what had previously been referred to as 'marine A and B clusters' of *Synechococcus*.

These cells are truly marine and have been isolated from both the coastal and the open ocean. All strains are obligate photoautrophs and are ca. 0.6-1.7 μm in diameter. This cluster is however further divided into a population that either contains or does not contain phycoerythrin. The reference strains are WH8103 for the phycoerythrin containing strains and WH5701 for those strains that lack this pigment (Waterbury *et al.* 1986b). More recently Badger (Badger *et al.* 2002) proposed the division of the cyanobacteria into a a- and a a-subcluster based on the type of rbcL (large subunit of ribulose 1,5-bisphosphate carboxylase/oxygenase) found in these organisms. A-cyanobacteria were defined to contain a form IA, while a-cyanobacteria were defined to contain a form IB of this gene. In support for this division Badger analyses the phylogeny of carboxysomal proteins, which appear to support this division. Also, two particular bicarbonate transport systems appear to only be found in a-cyanobacteria, which lack carboxysomal carbonic anhydrases.

Ecology and Distribution

Synechococcus has been observed to occur at concentrations ranging between a few cells per ml to 106 cells per ml in virtually all regions of the oceanic euphotic zone except in samples from the McMurdo Sound and Ross Ice Shelf in Antarctica (Waterbury *et al.* 1986b). Cells are generally much more abundant in nutrient rich environments than in the oligotrophic ocean and prefer the upper well lit portion of the euphotic zone (Partensky *et al.* 1999a). *Synechococcus* has also been observed to occur at high abundances in environments with low salinities and/or low temperatures. Synechococcus is usually far outnumbered by *Prochlorococcus* in all environments, where they co-occur. Exceptions to this rule are areas of permanently enriched nutrients such as upwelling areas and coastal watersheds (Partensky *et al.* 1999a). In the nutrient deplete areas of the oceans, such as the central gyres, *Synechococcus* is apparently always present, although only at low concentrations ranging from a few to 4x103 cells ml-1 (Olson *et al.* 1990b, Blanchot *et al.* 1992, Campbell & Vaulot 1993, Li 1995, Blanchot & Rodier 1996). Vertically *Synechococcus* is usually relatively equitably distributed throughout the mixed layer and exhibits an affinity for the higher light regime.

Below the mixed layer cell concentrations rapidly decline. Vertical profiles are however strongly influenced by hydrologic conditions and can be very variable both seasonally and spatially. Overall *Synechococcus* abundance often parallels that of *Prochlorococcus* in the water column. In the Pacific HNLC (High Nutrient Low Chlorophyll) zone and in temperate open seas where stratification was recently established both profiles parallel each other and exhibit abundance maxima just about the SCM (Olson *et al.* 1990b, Li 1995, Landry *et al.* 1996).

The factors controlling the abundance of *Synechococcus* still remain poorly understood, especially considering that even in the most nutrient deplete regions of the central gyres, where cell abundances are often very low, population growth rates are often high and not very drastically limited (Partensky *et al.* 1999a). Factors such as grazing, viral mortality, genetic variability, grazing, light adaptation, temperature as well as nutrients are certainly involved, but remain to be investigated on a rigorous and global scale. Despite the uncertainties it has been suggested that there is at least a relationship between ambient nitrogen concentrations and *Synechococcus* abundance (Blanchot *et al.* 1992, Partensky *et al.* 1999a) and an inverse relationship to *Prochlorococcus* (Campbell & Vaulot 1993) in the upper euphotic zone, where light is not limiting. One environment where *Synechococcus* thrives particularly well are coastal plumes of major rivers (Paul *et al.* 2000, Wawrik *et al.* 2002, 2003, 2004a, b). Such plumes are coastally enriched with nutrients such as nitrate and phosphate, which drives large phytoplankton blooms. High productivity in coastal river plumes is often associated with large populations of *Synechococcus* and elevated form IA (cyanobacterial) *rbcL* mRNA.

It should also be noted that *Prochlorococcus* in thought to be at least 100 times more abundant than Synechococcus in warm oligotrophic waters (Partensky *et al.* 1999a). Assuming average cellular carbon concentrations it has thus been estimated that *Prochlorococcus* accounts for at least 22 times more carbon in these waters and may thus be of much greater significance to the global carbon cycle than *Synechococcus*.

Trichodesmium

Trichodesmium, also called *sea saw dust,* is a genus of filamentous cyanobacteria. They are found in nutrient poor tropical and subtropical ocean waters (particularly around Australia, where they were first described by Captain Cook). *Trichodesmium* fixes atmospheric nitrogen into ammonium, usable also for other organisms. While far from the only nitrogen fixing bacteria, they are among the most important of the marine varieties, and are being extensively studied for their role in nutrient cycling in the ocean. Unlike other nitrogen fixing bacteria, *Trichodesmium* does not have heterocysts, nor any other specialised cells for this task. Furthermore, nitrogen fixation peaks at mid-day, i.e. occurs during the same time as photosynthesis. Inhibitor studies even revealed that photosystem II activity is essential for nitrogen fixation in this organism.

All this may seem contradictory at first glance, because the enzyme responsible for nitrogen fixation, nitrogenase, is irreversibly inhibited by oxygen. However, *Trichodesmium* utilises photosynthesis for nitrogen fixation by carrying out the Mehler reaction, during which the oxygen produced by PS

II is reduced again after PS I. This regulation of photosynthesis for nitrogen fixation involves rapidly reversible coupling of their light-harvesting antenna, the phycobilisomes, with PS I and PS II. *Trichodesmium* forms blooms and provides substrate for many small oceanic organisms (bacteria, diatoms, dinoflagellates, protozoa, and copepods).

Chloroflexi

The Chloroflexi are a group of bacteria that produce energy through photosynthesis. They make up the bulk of the filamentous anoxygenic phototrophs (formerly known as green non-sulphur bacteria), though some are classified separately as the Thermomicrobia. They are named for their green pigment, usually found in photosynthetic bodies called chlorosomes.

Chloroflexi are typically filamentous, and can move about through bacterial gliding. They are facultatively aerobic, but do not produce oxygen during photosynthesis, and have a different method of carbon fixation (photoheterotrophy) than other photosynthetic bacteria. Phylogenetic trees indicate that they had a separate origin.

Green Sulphur Bacteria

The green sulphur bacteria (Chlorobiaceae) are a family of phototrophic bacteria. No other bacterial families are known to be closely related to them, and they are accordingly placed in their own phylum (Chlorobi). The phylum is most closely related to Bacteroidetes.

Green sulphur bacteria are generally non-motile (one species has a flagellum), and come in spheres, rods, and spirals. Their environment must be anaerobic (oxygen-free), and they need light to grow. They engage in photosynthesis, using bacteriochlorophylls *c*, *d*, and *e* in vesicles called chlorosomes attached to the membrane. They use sulphide ions as electron donor, and in the process the sulphide gets oxidised, producing globules of elemental sulphur outside the cell, which may then be further oxidised. (By contrast, the photosynthesis in plants uses water as electron donor and produces oxygen.)

A species of green sulphur bacteria has been found living near a black smoker off the coast of Mexico at a depth of 2,500 metres beneath the surface of the Pacific Ocean. At this depth, the bacteria, designated GSB1, lives off the dim glow of the thermal vent since no sunlight can penetrate to that depth.

Halobacteria

In taxonomy, the Halobacteria (also Halomebacteria) are a class of the Euryarchaeota, found in water saturated or nearly saturated with salt. They are also called halophiles, though this name is also used for other organisms which live in somewhat less concentrated salt water. They are common in most environments where large amounts of salt, moisture, and organic material are available. Large blooms appear reddish, from the pigment bacteriorhodopsin. This pigment is used to absorb light, which provides energy to create ATP. Halobacteria also possess a second pigment, halorhodopsin, which pumps in chloride ions in response to photons, creating a voltage gradient and assisting in the production of energy from light. The process is unrelated to other forms of photosynthesis involving electron transport; however, and halobacteria are incapable of fixing carbon from carbon dioxide.

Halobacteria can exist in salty environments because although they are aerobes they have a separate and different way of creating energy through photosynthesis. Parts of the membranes of halobacteria are purplish in colour. These parts conduct photosynthetic reactions with retinal pigment rather than chlorophyll. This allows them to create a proton gradient across the membrane of the cell which can be used to create ATP for their own use.

Halobacterium

The genus *Halobacterium* consists of several species of archaea with an obligate aerobic metabolism which require an environment with a high concentration of salt; many of their proteins will not function in low-salt environments. They grow on amino acids in their aerobic conditions. Their cell walls are also quite different from those of bacteria, as ordinary lipoprotein membranes fail in high salt concentrations. In shape, they may be either rods or cocci, and in colour, either red or purple. They reproduce using binary fission (by constriction), and are motile. Halobacterium grows best in a 42 degree Celsius environment.

The genome of an unspecified Halobacterium species has been sequenced and comprises 2,571,010 bp (base pairs) of DNA compiled into three circular strands: one large chromosome with 2,014,239 bp, and two smaller ones with 191,346 and 365,425 bp. This species, called Halobacterium sp. NRC-1, has been extensively used for post-genomic analysis. Halobacterium species can be found in the Great Salt Lake, the Dead Sea, Lake Magadi, and any other waters with high salt concentration. Purple Halobacterium species owe their colour to bacteriorhodopsin, a light-sensitive protein which provides chemical energy for the cell by using sunlight to pump protons out of the cell. The resulting proton gradient across the cell membrane is used to drive the synthesis of the energy carrier ATP. Thus, when these protons flow back in, they are used in the synthesis of ATP. The bacteriorhodopsin protein is chemically very similar to the light-detecting pigment rhodopsin, found in the vertebrate retina.

Genus Halobacterium:

- *Halobacterium cutirubrum*
- *Halobacterium denitrificans*
- *Halobacterium distributum*
- *Halobacterium halobium*
- *Halobacterium lacusprofundi*
- *Halobacterium mediterranei*
- *Halobacterium noricense*

- *Halobacterium pharaonis*
- *Halobacterium saccharovorum*
- *Halobacterium salinarium*
- *Halobacterium sodomense*
- *Halobacterium trapanicum*
- *Halobacterium vallismortis*
- *Halobacterium volcanii*

Heliobacteria

The heliobacteria are a small family of bacteria that produce energy through photosynthesis. The primary pigment involved is bacteriochlorophyll *g*, which is unique to the group and absorbs at different frequencies than other photosynthetic pigments, giving the heliobacteria their own environmental niche. Photosynthesis takes place at the cell membrane, which does not form folds or compartments as it does in many other groups.

RNA trees place the heliobacteria among the Firmicutes. Unlike most other members they do not have gram-positive stains, since their cell walls are thin, but they share the characteristic absence of an outer membrane. They are also similar in other respects, including the ability to form endospores. They are the only group related to the gram-positive bacteria that conduct photosynthesis.

Heliobacteria are photoheterotrophic, using energy from light or chemicals but relying exclusively on organic sources for carbon, and are exclusively anaerobic. Whereas most other photosynthetic bacteria are predominantly aquatic, heliobacteria have been found mostly in soils, especially rice paddies.

Picoeukaryotes

The least well-studied and understood component of the photosynthetic picoplankton is a diverse array of small eukaryotic algae collectively referred to as picoeukaryotes. These small algal cells may be the most abundant eukaryotes on earth and occur at concentrations between 10^2 to 10^4 cell ml^{-1} in the photic zone throughout the oceans (Diez *et al.* 2001). Their diversity, distribution as well as their ecology however remain largely unknown (Partensky *et al.* 1997). Much of the difficulty associated with studying these algae results from the lack of morphological variation when viewed under conventional light microscopy. Cells often merely appear as little green balls (Potter *et al.* 1997) and display few distinguishing features (Thomsen 1986, Simon *et al.* 1994, Caron *et al.* 1999). Often cells are referred to as LRGTs (Little Round Green Things). Epi-fluorescence microscopy yields little additional clues and cells appear as red-fluorescing spheres with no distinguishing features. Electron microscopy and cell culture may alleviate some of these limitations, but are too expensive, too difficult and too time consuming to be feasible as routine oceanographic techniques. Similarly flow cytometry can be used to estimate bulk cell abundance, but completely lacks phylogenetic resolution.

Despite limitations progress has been made in several ways over the past decade. At least three novel algal classes have been identified and newly described. Moestrup (Moestrup 1991) described the new class Pedinophyceae composed of the two green algae *Pedinomonas micron* and *Pedinomonas minor* in 1991.

Shortly thereafter *Pelagomonas calceolata* was first identified and described in the marine ultraplankton (Andersen *et al.* 1993). 18S rDNA sequence data placed this species at an unresolved position among other chromophytic algae illustrating at the time how little was known regarding the phylogenetic diversity of heterokont algae (yellow-green algae with flagella of unequal length) in the environment. Pelagophytes are now thought to be an important and prolific component of the autotrophic picoplankton.

Most recently the algal class Bollidophyceae, has been identified and described in cultures and the environment (Guillou *et al.* 1999a, Guillou *et al.* 1999b). Although the Bollidophyceae are not thought to be a major component of the picoplankton their discovery nonetheless illustrates that numerous novel lineages of picoeukaryotes may yet be identified.

To circumvent some of the limitations of traditional microscope-dependent techniques for the description of picoplankton communities, molecular techniques have been widely used. Using group specific 18S rDNA probes in order to probe for the prymnesiophyte fraction of an 18S rDNA PCR (Polymerase Chain Reaction) amplicon generated using universal eukaryotic 18S primers, Moon-van der Staay (Moon-van der Staay *et al.* 2000) found that this group accounted for less of the amplified DNA than would be expected based on pigment ratios determined by HPLC.

While these data should be interpreted with caution due to the biased nature of PCR, the same group also demonstrated the presence of several novel prymensiophyte lineages in their 18S rDNA clone libraries, which had no equivalent among cultured species. Using a similar PCR based approach in order to amplify and clone picoplankton rbcL gene sequences the diversity of picophytoplankton in the oligotrophic Gulf of Mexico has also been studied (Pichard *et al.* 1997b, Paul *et al.* 2000b). In the first study only five unique rbcL gene-sequences were recovered.

These sequences however demonstrated the presence of algae spanning the diversity of almost the entire form I clade. In the later study the presence of a diverse array of eukaryotic algae in a low salinity coastal plume was detected. Eukaryotic rbcL DNA sequences were related to prasinophytes, prymnesiophytes, diatoms and pelagophytes and shared between 85 and 99 per cent similarity with cultured representatives in GenBank. In addition this study illustrated how unknown levels of genetic diversity within the picoplankton may not only exist in the form of divergent and novel lineages, but also in the form of small micro-diverse clades of closely related sequences.

Several other studies have since also addressed the diversity of picoeukaryotes in the environment using 18S PCR based clone libraries (Diez *et al.* 2001, Moon-van der Staay *et al.* 2001, Vaulot *et al.* 2001). These studies further support the notion that the oceanic picoplankton (both heterotrophic as well as autotrophic) is genetically very diverse and yet contains a considerable number of undescribed lineages. In order to identify and characterise some of the classified and non-classified species in the environment FISH (Fluorescent In-Situ Hybridisation) and DGGE (Denaturant Gradient Gel Electrophoresis) in particular has been useful.

Using FISH several novel groups have been observed both in the Pacific Ocean and in coastal waters indicating their ubiquity (Vaulot *et al.* 2001). Two basal stramenopile lineages have been studied in field samples and enrichments in the northwestern Mediterranean Sea using FISH (Massana *et al.* 2002). Cells were 2-3μm in diameter and bacteriovorus. One particular lineage may have accounted for as much as 46 per cent of the heterotrophic flagellates suggesting the tremendous importance of this unclassified stramenopile as a bacterial grazer.

Purple Bacteria

Purple bacteria or purple photosynthetic bacteria are proteobacteria that are phototrophic, i.e. capable of producing energy through photosynthesis.

They are pigmented with bacteriochlorophyll *a* or *b*, together with various carotenoids. These give them colours ranging between purple, red, brown, and orange. Photosynthesis takes place at reaction centres on the cell membrane, which is folded into the cell to form sacs, tubes, or sheets, increasing the available surface area.

Like most other photosynthetic bacteria, purple bacteria do not produce oxygen, because the reducing agent involved in photosynthesis is not water. In some, called purple sulphur bacteria, it is either sulphide or elemental sulphur. The others, called purple non-sulphur bacteria (aka PNSB), typically use hydrogen although somè may use other compounds in small amounts.

At one point these were considered families, but RNA trees show the purple bacteria make up a variety of separate groups, each closer relatives of non-photosynthetic proteobacteria than one another. Purple non-sulphur bacteria are found among the alpha and beta subgroups, including:

Rhodospirillales	
Rhodospirillaceae	e.g. *Rhodospirillum*
Acetobacteraceae	e.g. *Rhodopila*
Rhizobiales	
Bradyrhizobiaceae	e.g. *Rhodopseudomonas*
Hyphomicrobiaceae	e.g. *Rhodomicrobium*
Rhodobiaceae	e.g. *Rhodobium*
Other families	
Rhodobacteraceae	e.g. *Rhodobacter*
Rhodocyclaceae	e.g. *Rhodocyclus*
Comamonadaceae	e.g. *Rhodoferax*

Purple sulphur bacteria are included among the gamma subgroup, and make up the order Chromatiales. The similarity between the photosynthetic machinery in these different lines indicates it had a common origin, either from some common ancestor or passed by lateral transfer.

Bibliography

Arora, D.K.: *Handbook of Applied Mycology,* Dekker, London, 1992.

Aslam, M.S.: *Handbook of Biology,* Campus Books, New Delhi, 2000.

Aslam, M.S.: *Immunobiology,* Campus Books, New Delhi, 2000.

Bay, E.G.: *Science: An Introduction,* Westview Press, Westview Special Series on Africa, London, 1982.

Board, R.G.: *A Modern Introduction to Food Microbiology,* Blackwell Scientific Publications, London, 1983.

Cherry, J.P.: *Protein Functionality in Foods,* American Chemical Society, Washington D.C., 1981.

Dumazedier, J.: *Towards a Society of Leisure,* Free Press, New York, 1967.

Duncan, W.A.: *Engendering School Learning: Science, Attitudes and Achievement among Girls and Boys in Botswana,* University of Stockholm, Stockholm, 1989.

Dyke, S .F.: *The Chemistry of Vitamins,* Interscience Publishers, New York, 1965.

Eley, G.: *Wild Fruits and Nuts,* E.P. Publishing Ltd., Yorkshire, 1976.

Gill, L.S.: *Science, Technology and Gender,* Polity Press in Association with the Open University, Cambridge, London, 1992.

Govindjee, Šesták Z. and Peters W.R.: *The Early History of Photosynthetica,* Comenius University, Bratislava, 2002.

Haisel, D. and Pospíšilová, J.: *The Effect of Water Stress on Photosynthetic Pigments Contents and Gas Exchange Parameters,* Comenius University, Bratislava, 2004.

Hudson, F.: *Development in Food Proteins,* Applied Science Publisher, London, 1982.

Kak, Subhash C.: *Indian Physics: Outline of Early History,* ArXiv, London, 2003.

Krause, M.V. and L.K. Mahann: *Food Nutrition and Diet Therapy,* W.B. Saunders Co., Philadelphia, 1979.

Malcolm, S.: *Science and Technology for People,* Tamil Nader State Council for Science and Technology, Madras, 1985.

Samba, Murthy K.V.: *Science and Technology in Medieval India,* Pratap, Agra, 1982.

Šesták, Z.: *Bibliography of Reviews and Methods of Photosynthesis,* Kluwer Academic Publishers, London, 1998.

Šesták, Z.: *Chlorophyll Fluorescence Kinetics Depends on Age of Leaves and Plants,* Kluwer Academic Publishers, London, 1999.

Sharma, P.V.: *History of Medicine in India,* Academic, New Delhi, 1992.

Singh, Harkishan: *Pharmaceutical Education,* Vallabh Prakashan, New Delhi, 1998.

Sponsors: *Proceedings of the Southeast Asian Seminar on Women and Science in Developing Countries,* Kovalevskaia Fund, New York, 1987.

Stolte-Heiskanen, V.: *Science: Role of Changing India,* ISSC, New Delhi, 1991.

Subbarayappa, B.V.: *Science and Technological Exchanges between India and Soviet Central Asia,* Educational, New Delhi, 1985.

Theresa, Montini: *Science: Technology and Human Values*, Oxford, New York, 1993.

Turnbull, David, and Helen Watson-Verran: *Handbook on Science, Technology and Society*, Sage, Beverly Hills, London, 1994.

Ulman, P., Èatský J. and Pospíšilová, J.: *Photosynthetic Traits in Wheat Grown under Decreased and Increased CO_2 Concentration after Transfer to Natural CO_2 Concentration*, Institute of Plant Molecular Biology, Èeské Budìjovice, 2000.

Vágner, M., Lipavská H. and Tichá I.: *Photosynthetic Pigments and Gas Exchange of in Vitro Grown Tobacco Plants as Affected by CO_2 Supply*, International Society of Photosynthesis, Montreal, 1999.

Valcke, R.: *Transgenic Pssu-ipt Tobacco under Biotic Stress*, CSIRO Publishing, Melbourne, 2001.

Van der Est, A. and Bruce, D.: *Photosynthesis: Fundamental Aspects to Global Perspectives*, International Society of Photosynthesis, Montreal, 2005.

Veldhnis, M.K.: *Food Science and Technology*, The AVI Publishing Co., Connecticut, 1977.

Vièánková, A., Kutík J., Holá D., Koèová M. and Wilhelmová N.: *The Chloroplast Ultrastructure and Photosynthetic Characteristics of Maize Leaves in Various Parts of the Leaf Blade and during the Diurnal Cycle*, Institute of Plant Molecular Biology, Èeské Budìjovice, 2001.

Vijaya Lakshmi, Sharma: *Science in Pre-Modern India*, D.K. Print World, New Delhi, 2001.

Vomáèka, L. and Pospíšilová J.: *Rehydration of Sugar Beet Plants after Water Stress: Effects of Cytokinins*, Institute of Plant Molecular Biology, Èeské Budìjovice, 2003.

Vomáèka, L.: *Gas Exchange and Cytokinins*, CSIRO Publishing, Melbourne 2001.

Wardha, B.: *Science and Technology in India*, Centre of Science for Villages, Wardha, 1983.

Whitaker, J.R.: *Principles of Enzymology for Food Science*, Marcel Dekker, New York, 1977.

Winslow, A.: *Science: History*, Duke University Press, London, 1990.

Zima, M.: *Ecophysiology of Plant Production Processes in Stress Conditions*, Slovak Agr. Univ., Nitra, 2000.

Zimmerman, J.: *Science in India: Today and Tomorrow*, Praeger Publishers, London, 1983.

Index

P

Q

R

S

T

V

W

X

Z

□□□